# Praise for *The Cyber Security Handbook*

'Stephen and Sébastien are rock stars on cyber security and rehearsing for cyber attacks. Plus, Sébastien brings his Krav Maga's advanced experience to boost his education. This is unique and SO powerful.'

**Rémi Lassiaille, Vice President, IBM**

'Whether you come from cyber or crisis communications or don't have a clue about what it is, *The Cyber Security Handbook* is a must to start AND deepen your knowledge. It should be mandatory for future generations in comms.'

**Carla Portier, VP Technology , PR Company**

'Forget theory, this book dives straight into practical insights, gripping real-life case studies, and actionable tips for tackling cyber security. It's a game-changer that also shows how to build stronger business resilience. I'll be reaching for it again and again.'

**Rina Singh, Head of Business Resilience,**
**Data Communications Company**

'The realistic scenarios and guidance will significantly enhance your cyber crisis management capabilities, fostering crucial discussions among high-level executives.'

**Marcelo Nicácio, CIO, Tredegar Corporation**

'You think you are safe? You are not. Perhaps a little more after having read this book.'

**Raphael Sanchez, President AND CEO, Generix Group**

'Prepare for the digital storms before they strike. This book is your survival guide to navigating and mastering cyber crises.'

**Patrice Bordron, SVP, Chief Digital Information Officer,**
**Community Health Systems**

'If you're looking to understand what to prioritise to best protect your company from cyber risks, this is the real deal: field tested best practices and concrete steps to take and compelling case studies. The practical guide to amp up your cyber game!'

**Nicolas Meyerhoffer, Vice-President,**
**Managing Director Financial Services, IBM France**

'The cyber threat has never been more complex. Companies must take this threat very seriously and anticipate it well in advance of their project, which is why I recommend this book.'

**Yann Bonnet, deputy CEO of Campus Cyber**

'A fantastic resource that highlights the importance of preparation and practice in building cyber resilience, far beyond just technical measures. Packed with relevant case studies and examples *The Cyber Security Handbook* serves as a great primer for executives.'

**Adam Harrison, Managing Director, consultancy firm**

# The Cyber Security Handbook

# Pearson

At Pearson, we believe in learning – all kinds of learning for all kinds of people. Whether it's at home, in the classroom or in the workplace, learning is the key to improving our life chances.

That's why we're working with leading authors to bring you the latest thinking and best practices, so you can get better at the things that are important to you. You can learn on the page or on the move, and with content that's always crafted to help you understand quickly and apply what you've learned.

If you want to upgrade your personal skills or accelerate your career, become a more effective leader or more powerful communicator, discover new opportunities or simply find more inspiration, we can help you make progress in your work and life.

Every day our work helps learning flourish, and wherever learning flourishes, so do people.

To learn more, please visit us at **www.pearson.com**

## The Financial Times

With a worldwide network of highly respected journalists, *The Financial Times* provides global business news, insightful opinion and expert analysis of business, finance and politics. With over 500 journalists reporting from 50 countries worldwide, our in-depth coverage of international news is objectively reported and analysed from an independent, global perspective.

To find out more, visit **www.ft.com**

Stephen Delahunty and
Sébastien Jardin

# The Cyber
# Security
# Handbook

How to protect your business from
cyber threats and attacks

Translation edited by Ruth Simpson

# Pearson

Harlow, England • London • New York • Boston • San Francisco • Toronto • Sydney • Dubai • Singapore • Hong Kong
Tokyo • Seoul • Taipei • New Delhi • Cape Town • São Paulo • Mexico City • Madrid • Amsterdam • Munich • Paris • Milan

**PEARSON EDUCATION LIMITED**
KAO Two
KAO Park
Harlow CM17 9NA
United Kingdom
Tel: +44 (0)1279 623623
Web: www.pearson.com

**First edition published 2025** (print and electronic)
© Pearson Education Limited 2025 (print and electronic)

Authorised translation from the French language edition entitled *Crise cyber: Se préparer au pire pour donner le meilleur*, ISBN 9782744068706, by authors Stephen Delahunty and Sébastien Jardin, published by Pearson France @Copyright_2024.

The rights of Stephen Delahunty and Sébastien Jardin to be identified as authors of this work have been asserted by them in accordance with the Copyright, Designs and Patents Act 1988.

The print publication is protected by copyright. Prior to any prohibited reproduction, storage in a retrieval system, distribution or transmission in any form or by any means, electronic, mechanical, recording or otherwise, permission should be obtained from the publisher or, where applicable, a licence permitting restricted copying in the United Kingdom should be obtained from the Copyright Licensing Agency Ltd, Barnard's Inn, 86 Fetter Lane, London EC4A 1EN.

The ePublication is protected by copyright and must not be copied, reproduced, transferred, distributed, leased, licensed or publicly performed or used in any way except as specifically permitted in writing by the publishers, as allowed under the terms and conditions under which it was purchased, or as strictly permitted by applicable copyright law. Any unauthorised distribution or use of this text may be a direct infringement of the authors' and the publisher's rights and those responsible may be liable in law accordingly.

All rights reserved. No part of this book may be reproduced or transmitted in any form or by any means, electronic or mechanical, including photocopying, recording or by any information storage retrieval system, without permission from Pearson Education, Inc. English language electronic edition published by Pearson Education Limited, © 2025.

We may use artificial intelligence (AI) in our products and services, including to support the creation of content.

All trademarks used herein are the property of their respective owners. The use of any trademark in this text does not vest in the author or publisher any trademark ownership rights in such trademarks, nor does the use of such trademarks imply any affiliation with or endorsement of this book by such owners.

Pearson Education is not responsible for the content of third-party internet sites.

ISBN: 978-1-292-74747-7 (print)
      978-1-292-47956-9 (ePub)

**British Library Cataloguing-in-Publication Data**
A catalogue record for the print edition is available from the British Library

**Library of Congress Cataloging-in-Publication Data**
A catalog record for the print edition is available from the Library of Congress

10 9 8 7 6 5 4 3 2 1
29 28 27 26 25

Cover design by Michelle Morgan
Cover image S and V Design/iStock/Getty Images

Print edition typeset in 9/13 Stone Serif ITC Pro by Straive
Printed and bound by CPI UK

NOTE THAT ANY PAGE CROSS REFERENCES REFER TO THE PRINT EDITION

'Know the enemy,

Know yourself,

And victory

Is never in doubt,

Not in a hundred battles.'

– Sun-Tzu, *The Art of War*

While we work hard to present unbiased, fully accessible content, we want to hear from you about any concerns or needs with this Pearson product so that we can investigate and address them:

- Please contact us with concerns about any potential bias at https://www. pearson.com/report-bias.html

- For accessibility-related issues, such as using assistive technology with Pearson products, alternative text requests, or accessibility documentation, email the Pearson Disability Support team at disability.support@pearson.com

# Contents

# List of case studies

# Table of acronyms

| | |
|---|---|
| AI | Artificial Intelligence |
| ANSSI | *Agence nationale de la sécurité des systèmes d'information* (France) |
| APT | Advanced Persistent Threat |
| BEC | Business Email Compromise |
| CDEI | Centre for Data Ethics and Innovation (UK) |
| CDN | Content Delivery Networks |
| CERT | Computer Emergency Response |
| CIO | Chief Information Officer |
| CISO | Chief Information Security Officer |
| CSP | Cloud Service Provider |
| CTI | Cyber Threat Intelligence |
| DCMS | Department for Digital, Culture, Media and Sport (UK) |
| DDoS | Distributed Denial-of-Service |
| DORA | Digital Operational Resilience Act (EU) |
| DSTL | Defence Science and Technology Laboratory (UK) |
| ENISA | European Union Agency for Cybersecurity (EU) |
| FCA | Financial Conduct Authority (UK) |
| GAN | Generative Adversarial Network |
| GCHQ | Government Communications Headquarters (UK) |
| GDPR | General Data Protection Regulation (EU) |
| IaaS | Infrastructure as a Service |
| IAB | Initial Access Broker |
| ICO | Information Commissioner's Office (UK) |
| ICT | Information and Communication Technology |
| IoT | Internet of Things |
| IS | Information System |
| ISM | Information Systems Management |
| ISP | Internet Service Provider |
| IT | Information Technology |
| ITIL | IT Infrastructure Library |
| MFA | Muti-factor Authentication |
| MoD | Ministry of Defence (UK) |

| NIS | Network and Information Systems |
|---|---|
| NSA | National Security Agency (US) |
| NCSC | National Cyber Security Centre (UK) |
| OEM | Original Equipment Manufacturer |
| OODA (model) | Observe, Orient, Decide, Act |
| OT | Operational Technology |
| PaaS | Platform as a Service |
| PACE (model) | Primary, Alternate, Contingency, and Emergency |
| PII | Personally Identifiable Information |
| PRA | Prudential Regulation Authority (UK) |
| QKD | Quantum Key Distribution |
| RaaS | Ransomware as a Service |
| RBAC | Role-Based Access Control |
| RPA (model) | Repetition, Precision, Anchoring |
| SaaS | Software as a Service |
| SIEM | Security Information and Event Management |
| SLA | Service Level Agreement |
| SMEs | Small and Medium Enterprises |
| SOC | Security Operations Centre |
| UEBA | User and Entity Behaviour Analytics |

# About the authors

**From the TV Control Room to the Krav Maga Mat: The Unlikely Team Shaping Cyber Resilience**

**Stephen Delahunty** is a certified member of the Business Continuity Institute (CBCI) with significant expertise in cyber resilience and crisis management. After 15 years in television, working with industry leaders like Sony and NBC, he moved from directing live broadcasts to directing live-fire drills, applying the same calm leadership and decision-making to prepare organisations for high-pressure scenarios.

© Kim Hardy

At IBM Cyber Range, Stephen worked as Lead Facilitator, where he collaborated with clients and colleagues like Sébastien to develop innovative, immersive training programmes for managing cyber crises. His efforts helped organisations strengthen their preparedness and build confidence in facing complex challenges.

Now, as Global Training Lead for Cognizant's Integrated Crisis Management team, Stephen focuses on ensuring readiness for crises of all kinds, from cyber attacks to extreme weather events. His work enables teams to respond effectively to disruptions while embedding resilience into everyday operations.

From the TV control room to the boardroom, Stephen has built a reputation for turning high-stakes business challenges into opportunities for growth and resilience.

**Sébastien Jardin** has spent his career preparing organisations for moments they hope will never come. As a cyber crisis management coach and strategic consultant, he works with leaders to build the clarity and confidence needed to handle the most challenging simulations. His approach goes beyond standard training, offering highly tailored, scenario-driven exercises that refine leadership decision-making at critical moments.

At IBM Cyber Range, Sébastien set new standards for immersive crisis simulations, shaping how organisations approach resilience and leadership under pressure. Today, as Director of Cyber Resilience at Deloitte, he partners with leadership teams across Europe, Africa and North America to position cyber security as a strategic priority rather than a back-office concern.

A two-time winner of the French *"Trophées de la Sécurité"* awards, Sébastien's expertise spans more than 15 years in security and complex sales. A certified Krav Maga instructor, he applies realism and precision to his methods, working with elite French Army operatives and law enforcement to ensure his strategies are as practical as they are effective.

When he's not designing training programmes or advising stakeholders, Sébastien shares his insights as a columnist for the French edition of *Harvard Business Review* (HBR). Whether speaking at conferences or collaborating with leadership teams, he brings a unique combination of technical expertise, communication skills, and strategic insight to every challenge, helping organisations prepare for what's next.

# Authors' acknowledgements

**From Stephen**

I would like to thank our commissioning editor, Eloise Cook, and the Financial Times Publishing team for their support. I also wish to give a special thank you to Delphine Galtier, who commissioned the original French edition, making all this possible. To Ruth Simpson, our translator, *merci beaucoup*!

Thanks to Laurance Dine, who recognised the value that a creative from the media sector could bring to the cyber security industry. Thank you to Jake Paulson and my former colleagues at the IBM X-Force Cyber Range! And to all the interviewees whose insights have greatly enriched this book.

**From Sébastien**

I was able to write this book by drawing on conversations I had with some talented people over the course of my career. I would like to thank them all, and in no particular order:

Thierry Libaert, former Head of Crisis Communication at the EDF Group, advisor to the European Economic and Social Committee (EESC), scientific collaborator at the Earth and Life Institute, former Professor at the Catholic University of Louvain, Associate Lecturer at Celsa-Sorbonne and Lecturer at Sciences Po Paris. Thank you, Thierry, for your friendship over the last fifteen years, for the hours spent reading your communication books, for your supervision of my IAE dissertation on crisis communication, for our Parisian lunches and for our conversations about music.

Shamla Naidoo, former IBM Global CISO, Faculty member of IANS, Professor at the Chicago School of Law, Senior Advisor & Board Member, Director of Strategy and Innovation at Netskope. Thank you, Shamla, for taking on this crazy project in France, for your openness and kindness, for your boundless creative spirit, and for all the time you spent explaining the details of life as a CISO (*Chief Information Security Officer*).

Marianne Le Huu, former CIO at Indosuez, psycho-behavioural naturopath and my first boss at IBM. Thank you, Marianne, for believing in a young man from the south of France and for throwing me into the world of complex project sales without even asking me. Thank you for your trust, your smile, your kindness and your concern for others.

Daniel J. W. King, former US Army and Liaison for the US Cyber Command, my first boss at the IBM Security Cyber Range in Cambridge, lecturer at Endicott College, Chief of Cybersecurity for CISA in Massachusetts. Thank you, Daniel, for opening the door to the Cyber Range and giving me the opportunity to explore the role of cyber crisis instructor for senior executives.

Thierry Delville, VP Group CISO of Capgemini, former partner at PWC France, former Head of the Internal Security Technology department, former Director of Technical and Logistical Services at the Paris police headquarters, former ministerial delegate for security industries and the fight against cyber threats. Thank you, Thierry, for your attentive ear during your leadership of your final unit in the French Security Administration, for welcoming me with open arms and your natural kindness when our paths crossed again a few years later in the cyber world. Thank you also for your comments on French villages and your holistic vision of the world of security.

Rodrigue Le Bayon, Head of Global CERT at Orange Cyberdefense, who I met when he was in charge of the CyberSOC division at Orange Cyberdefense. Thank you, Rodrigue, for your welcome to the cyber world, your confidence, your positivity, your open-mindedness, and that day at SOC when I was your guest.

Lee Morrison, founder of the Urban Combatives Group, who I had the good fortune to meet a few years ago on a personal protection course and who confirmed both my view on the need to train realistically ("Train like you fight"), but also agreed that safety is above all a mindset.

Special thanks to Delphine Galtier at Pearson France who made this book a reality a year ago, and to Eloise Cook and the Pearson UK team for turning it into an international reality.

# Foreword

**James Lodge FBCI, MBCP**
**Global Continuity Manager, Multinational Law Firm**

Cyber dangers have surged in recent years, upsetting society and economies all over the globe, just as changing digital tides do. Regular news of terrible assaults, data breaches and digital disruptions takes a subtle but major toll on both people and businesses. Many struggle with the direct and indirect effects of cyber events influencing their colleagues, businesses and communities; circumstances that sometimes seem outside of their control.

The writers of *The Cyber Security Handbook* acknowledge this reality beautifully. More than simply another technical manual, what Stephen Delahunty and Sébastien Jardin have produced is a thorough guide for negotiating the turbulent waters of modern cyber security. Their work honours the technological complexity as well as the very personal aspects influencing our digital resilience.

Having seen the development of cyber security over many years, I find it amazing how this handbook closes important knowledge gaps. It addresses technical teams as well as board members since real cyber resilience results from an organisation-wide security commitment. Combining Delahunty's experience in television production and crisis management with Jardin's knowledge of cyber security, the writers provide original insights on getting ready for and handling cyberattacks.

The timing of this work could not be more relevant. At a turning point in our digital path, artificial intelligence, quantum computing and linked systems present unheard-of opportunities coupled with equally unprecedented risks. The writers know that in this environment cyber security is about safeguarding the trust and resilient underpinnings of our digital society, not only about data or systems.

This handbook distinguishes itself by realising that, fundamentally, cyber security is a human challenge. Although technical answers are very important, the writers stress that our reaction to cyber attacks has to start with people. From the server room to the boardroom, they walk readers through the key stages of developing resilience at every level of an organisation.

Especially remarkable is the way the guidebook addresses crisis readiness. Based on practical knowledge, the writers offer a cyber crisis management model that addresses the emotional as well as the technical aspects of reaction. They are aware that during a cyber attack companies have to negotiate not just the immediate technical difficulties but also the knock-on repercussions affecting staff, consumers and investors.

Their investigation of the cyber criminal ecology is similarly instructive. Instead of depicting attackers as dark people in hoodies, a caricature the writers aggressively resist, they offer a complex picture of the contemporary cyberspace. This realisation helps readers to better grasp their enemies and, so, to strengthen their defences.

One should pay particular attention to the way the writers handle communication during cyber emergencies. Their advice on crisis communication is extremely helpful in a time when one tweet may affect reputation and share values. They provide sensible models for preserving openness and confidence while handling the difficult technological problems of a cyber crime.

Most crucially, this handbook notes that cyber security is a journey rather than a destination. The writers know that companies in our fast-changing digital terrain have to develop flexibility in addition to technological knowledge. They stress the need of ongoing education, frequent testing, and the development of strong response muscles by means of useful exercises.

The sections on employee preparation show a great awareness of organisational behaviour and human psychology. The writers offer a good foundation for involving staff members as active participants in cyber defence, instead of considering them as possible weaknesses. Their method acknowledges that each person working for a company is absolutely essential in preserving digital resilience.

The way the handbook treats senior management preparation is especially sophisticated. The authors understand that leaders must not only comprehend technical risks but also be prepared to make critical decisions under pressure. Their guidance on executive training and crisis simulation provides practical tools for building this essential capability.

The writers strike a careful balance between technical accuracy and understandable explanation all through the work. They understand that good cyber security calls for both wide knowledge and great depth of skill. The handbook is helpful for readers at all technical levels since it shows how to communicate difficult ideas without oversimplification.

The way the writers tackle issues, problems and new technologies that are likely to be problematic in the future shows great foresight, and they offer frameworks for recognising and adjusting to these shifting risks in our fast-changing digital environment.

This forward-looking approach helps readers prepare not just for today's known challenges but for tomorrow's uncertainties. By focusing on adaptable strategies rather than fixed solutions, they equip organisations to face an ever-evolving threat landscape with confidence and resilience.

The handbook's focus on developing cyber resilience via frequent testing and simulation reveals a great awareness of how companies grow and change. Although academic knowledge is valuable, they understand that practical experience has to complement it. Their advice on creating successful cyber exercises gives companies instruments to develop real reaction capabilities.

Especially perceptive is their investigation of the link between IT departments and corporate executives. The writers know that good cyber security calls for breaking down barriers and encouraging real teamwork across organisational lines. Their pragmatic guidance on creating these connections tackles one of the most often occurring issues in corporate cybersecurity.

The sections on supply chain security and third-party risk management capture the ever-linked character of modern business. The writers understand that organisations have to develop real resilience outside of their own internal boundaries. Their advice on handling these intricate connections offers useful instruments to handle one of the most urgent cyber security issues of today.

The writers of the handbook have a measured, optimistic attitude throughout, and although they never minimise the gravity of cyber attacks, they constantly stress that companies can create significant resilience by methodically planning and practicing.

The way that they look at and deal with regulatory compliance and legal obligations is particularly well handled. They aim to illustrate how these criteria may be adopted into a broad security plan instead of treating them as just checkboxes and marking them off before them move on. This approach will enable companies to go beyond compliance to actual security capabilities.

The way the handbook addresses incident response and recovery offers insightful analysis for companies of all security degrees of sophistication. The writers know that not every attack can be stopped, hence their advice on developing strong response capacity helps companies be ready for unavoidable difficulties.

Their investigation of the human effects of cyber events is highly valuable. Behind every hack or attack, the writers understand, are people coping with stress, uncertainty, and often dread. Their advice on helping teams overcome obstacles keeps companies both operational and human resilient.

Including thorough response systems and communication templates gives companies useful tools fit for their particular requirements. These tools, based on actual experience, enable readers to move from understanding to action.

Looking ahead, the writers offer insightful analysis of growing concerns and changing security issues, which will allow companies to create flexible security systems capable of developing alongside shifting risks.

Artificial Intelligence and machine learning in cyber security is something that can't be ignored and is one of the moat discussed topics in recent times, and especially relevant in a world that is moving at a fast pace in this field. The handbook attempts to cut through the hype to offer sensible advice on using these technologies while managing and controlling their inherent risks.

Throughout this piece, the writers know that real security comes from developing an organisational capacity to react to changing risks rather than from any one specific tool or approach. It's fair to say that modern security initiatives can benefit significantly from this level of focus on resilience placing it above simple protection.

As such their advice on raising security awareness and developing culture goes a long way to showing a deep and thorough knowledge of organisational transformation. Instead of offering awareness as a set of basic information or course style instruction, they offer actual models for cultivating real security consciousness throughout businesses.

The handbook's treatment of crisis communication and stakeholder management is especially valuable in today's connected world. The writers realise that controlling perception is sometimes more important than controlling technical events, hence their advice guides companies across these challenging waters.

The ideas and direction this handbook offer become ever more valuable as we look towards an ever more digital future. The writers have developed a road map for organisational resilience in our digital age rather than only a guide for cyber security.

Finally, this handbook is absolutely vital for our knowledge of organisational resilience and cyber security. Stephen Delahunty and Sébastien Jardin have created work that will enable companies to negotiate the difficult terrain of our digital future. Their pragmatic knowledge, based on actual experience and improved by new ideas, offers great direction for everyone working towards actual cyber resilience.

Our common dedication to security and resilience is perhaps our most unwavering anchor in these turbulent digital times. This handbook guides us in enhancing that anchor by knowledge, preparation and ongoing development. For anyone in charge of guiding their company across the difficult waters of our digital age, it is indispensable reading.

# Preface

**Stephen Delahunty**

In 2021, I packed my bags for Boston to meet the IBM Cyber Range team. As part of IBM's X-Force division, I was no stranger to cyber security, but the Cyber Range – IBM's crown jewel for client training – was something else entirely. As an Irishman, Boston carried its own significance. Known as the unofficial capital of the Irish diaspora and steeped in history, it felt like the perfect place to start this new chapter.

The trip didn't disappoint. Boston was everything I hoped it would be, and the Cyber Range team was just as impressive. It was there I met Sébastien. I watched him lead a cyber crisis training session and was immediately impressed by his approach. As it turned out, we'd both been vying for the same role. Cue the awkward handshake! Thankfully, IBM hired us both, we became fast friends and quickly realised how well our different backgrounds complemented each other professionally.

IBM had launched the Cyber Range in 2016 but, by 2021, the landscape had shifted. Hybrid workforces, increasingly sophisticated threats, and mounting pressure on senior leaders meant training had to evolve. This wasn't just a matter of refining skills; it required rethinking the format entirely. Luckily, Sébastien and I brought a unique mix of perspectives. His Krav Maga expertise taught reflexive decision-making under pressure, while my 15 years in television production focused on creating experiences that resonate. Together, we reimagined the Cyber Range's approach to training: immersive, flexible, and – most importantly – aligned with the real challenges our clients faced.

Our mission was clear. CISOs and CIOs were fighting battles on two fronts: defending their organisations against increasingly sophisticated hackers while also trying to convince leadership to prioritise cyber security. We aimed to bridge that gap. "We'll run a crisis simulation," we'd say, "and by the end, your executives won't just understand cyber security – they'll champion it."

One session in Paris stands out. We were working with the executive team of a major international hotel chain on a ransomware exercise. Thanks to a last-minute reshuffle, I was leading the session while Sébastien worked behind the scenes. The room was filled with the sound of the executives settling in. Then the CEO's voice could be heard over the noise.

"What's the point of all this?" he asked, cutting straight to the chase. "Why not just accept ransom payments as a cost of doing business?"

It was a fair question – and a sharp one. Scepticism often kicks off these sessions, and it's part of what makes them so engaging. I paused, and replied, "Great question. But this isn't about avoiding attacks – it's about surviving them."

I shared a story from a trip to Rio de Janeiro, where my hosts advised me to carry two wallets: one for myself and one for the thief. "The idea," I explained, "wasn't to stop the thief but to make sure I walked away intact. Crisis simulations are like that second wallet. They prepare you to survive an attack, minimise damage, and recover stronger."

The analogy landed. The CEO nodded, the energy in the room shifted, and the simulation began. Over the next few hours, the team tackled escalating scenarios, testing their decision-making and resilience. By the end, the scepticism was gone. The CEO's initial challenge had sparked the urgency the session needed, and the team left with a clear understanding of what it takes to protect both their people and their business.

We were invited back six months later. That Paris session wasn't just a successful training – it was a reminder of why we do this work. Cyber resilience isn't just about technology or policies. It's about preparing people to lead through chaos and come out the other side stronger.

**Sébastien Jardin**

When I entered the world of cyber security in 2017, I braced myself for an environment that would be deeply technical – and I wasn't wrong. Acronyms were everywhere, technological imperatives ruled, and the quirks of IT culture were unmistakable. It felt like stepping into a parallel universe, complete with its own language, rituals, and a distinctive sense of style.

But it didn't take long to realise that beneath the technical surface lay something more familiar. Through conversations with colleagues, partners, and clients across Europe and the United States, I found that while cyber security is undeniably a haven for highly skilled experts, a significant part of the job falls squarely within the realms of communication and leadership. In particular, the Chief Information Security Officers (CISOs) I worked with were navigating a complex role: they were not just safeguarding systems, but also championing security within their organisations and rallying others to their cause.

Almost every CISO I met shared a similar struggle. They faced a rising tide of threats but lacked the financial and human resources to address them effectively. Many expressed frustrations at the challenge of mobilising their teams and playing their critical role as the first line of defence. Above all, they saw a need to bring cyber security out of its bubble of technical experts and into the boardroom, where it could be recognised as the strategic issue it truly is.

This realisation shaped my journey in cyber security. After roles in cyber sales and team leadership, I found myself once again blending my professional and personal passions. My years of Krav Maga training – teaching reflexes and decision-making under pressure – proved unexpectedly valuable in helping CISOs and their leadership teams prepare for cyber crises. My goal was clear: sharpen their instincts and empower them to respond effectively to high-stakes situations while ensuring they secured the budgets and support they needed to protect their organisations.

This book isn't a technical manual for managing cyber crises. Instead, it offers insights, stories, and lessons from the field, showing that even the most talented technical teams can't succeed without management and staff fully on board. Without proper preparation, responses to crises risk being improvised and chaotic at best – or catastrophic at worst.

Rather than instilling fear, this handbook aims to inspire awareness and action. Effective cyber crisis management requires regular, realistic training that prepares every part of the organisation – IT teams, employees, and leadership alike. Why is this preparation so vital? How should you train, and how often? These are just some of the questions we'll explore together.

This book also marks the result of a wonderful partnership with my colleague and friend from Ireland, Stephen Delahunty. We met in Boston in 2021 and quickly discovered a shared vision for helping organisations navigate cyber crises. A sharp mind, sharper skills, and a hell of a sense of humour, that one! Our collaboration culminated in a moment of inspiration one evening in Paris: "Why don't we write a book about our experiences?" A few months later, here we are.

This book is intended to be a practical guide to cyber crisis management but also an enjoyable read . . . let us know if we hit the mark!

# INTRODUCTION

> "Truth is by nature self-evident.
> As soon as you remove the cobwebs of
> ignorance that surround it, it shines clear."
>
> – Gandhi

Let's start with a time out. Take a break from your hectic schedule and look around the room you are in. How many digital or connected devices are there? Your mobile phone, your smart meter from British Gas, the thermostat from Hive, or even the Nespresso machine you programmed last night, whistling away as it gets your caffeine fix ready. Now think about your entire home, your office, production lines, research laboratories, operating theatres, planes, trains. . . and imagine all the digital or connected devices there too. Technology is everywhere. And that's a good thing, because it can be harnessed to improve performance, reduce pollution, optimise resources, and make everyday life easier for billions of people around the world. And if industry specialists are to be believed, the digital wave is growing.

But like so many innovations throughout history, it is both wonderful and devastating. While working at Bletchley Park during World War II, the codebreakers there could never have imagined how their ground-breaking work in cryptography would one day lead to the sophisticated encryption methods we use today. Yet those very methods can also enable the concealment of cyber attacks and crimes on a massive scale.

Here's what James Hatch, Director of Cyber Security at BAE Systems, has to say about the evolving landscape:

"The digital world has grown from niche expertise into an essential part of everyday life. In sectors like defence and finance, data integrity, confidentiality, and resilience have always been paramount, as security failures in these areas can have catastrophic effects. But the pace of technological growth has made digital security an urgent concern for everyone, not just specialists. The internet, and the systems built upon it, were never designed for the level of threats we now face today."[1]

---

[1] Hatch, J., "Defending the digital world: The growing cyber threats", BAE Systems, https://www.baesystems.com/en/digital/blog/defending-the-digital-world, 26 June 2017.

To reinforce this, Ian Levy, Technical Director at the UK's National Cyber Security Centre (NCSC), offers a perspective from a national security viewpoint:

> *"People often think of cyber security as a technical issue, but it's really about behaviour. Most of the large-scale breaches we see aren't due to sophisticated hacking; they result from poor security hygiene, like weak passwords or unpatched software. The rise in connected devices has only multiplied the number of entry points for attackers. Our priority must be to ensure that individuals and organisations are well-prepared to respond quickly and effectively when breaches occur."[2]*

People have been learning how to use applications and software packages for over 30 years, but has anyone taught them how computers really work? How to become IT specialists? No, and that's good, because computer users are probably the least receptive to safety instructions. The important thing is the attitude that users adopt when dealing with events.

That's what wargames are all about: training individuals and those in command (management) to deal with unforeseen situations.

Cyber criminals, governments, and hacktivist groups have understood that perfectly. Seeing more and more digital attacks every year, and with ever greater pressure to deliver products and services quickly and at the best price, they are constantly honing their malevolent actions for their own financial, policy, or ideological ends.

Each week, another public or private company makes headlines for being the target of a cyber attack. But let's be clear, those reports are just the tip of the iceberg. They don't cover minor incidents or attempted attacks. They only deal with cyber crises that have a serious impact on operations. And nobody is immune. Large or small, public or private. Everyone is affected or, to quote Jean de La Fontaine: "They did not all die, but they were all struck down."

Financial losses can be huge. In the UK, the TalkTalk hack in 2015 cost the company £60 million and led to a £400,000 fine from the Information Commissioner's Office (ICO). Sometimes senior management must resign because the losses are so severe, and the handling of the crisis so catastrophic. For instance, in the British Airways data breach of 2018, a fine of £183 million was initially proposed by the ICO after 400,000 customer records were compromised. Sometimes, lives are lost: a patient died in the 2021 cyber attack on Düsseldorf University Hospital in Germany.

The survival of a business, or even an entire economy, might be at stake. In the UK, 99.9% of businesses are SMEs, and according to insurers, the risk of

---

[2] Levy, I., "Cybersecurity and behaviour: Beyond technical defences", NCSC.

an SME going bankrupt increases by 50% in the six months following a cyber attack. These examples show that cyber security, i.e., risk mitigation for digital technology, vastly exceeds technical issues.

If we are to believe the excellent book *Cyberattaques*[3] by our colleague Gérôme Billois, cyber threats are nothing new, dating back to the very origins of the internet. Over time, they have evolved into a harsh reality, with cyber crime costing the UK economy an estimated £27 billion annually,[4] according to a report by the UK government. To put that in perspective, it's the equivalent of multiple economic crises. Remember the 2007 financial crash? Cyber crime could easily cause similar disruption, and it's on the rise. Globally, the cost of cyber crime is projected to reach a staggering $10.5 trillion by 2025.[5] It's a real and highly lucrative business, much less risky than firing a rocket launcher at a cash courier in the street. If Pablo Escobar were alive today, he might well have traded his cocaine for a computer.

Our lives revolve around digital technology. Can you imagine an office job without IT? And how would you react if you were held hostage by someone who had taken control of your devices? Just take a second and imagine what would happen if someone hacked your father's pacemaker and you had to pay to stop him from jumping up and down on the bed rather than sitting peacefully in his chair. People often ask if ransoms should be paid. Of course they shouldn't, for many reasons, but when you take a close look at the figures, people don't always follow that advice.

So what should we do? Give up, set aside a large sum of money in cryptocurrency, and brace ourselves? Or stand tall, shoulders straight, and realise that when it comes to cyber crises, it's not a question of if, but when, and start getting ready? Make sure everyone in the organisation is playing their role as the first line of defence, so that IT and cyber-security teams stay up to date, and senior management can make the right decisions under pressure.

This book provides answers to all these questions, and much more. Make it your bedside guide, a daily reminder of healthy digital habits. It will also open your eyes to just how dependent we are on digital technology and how we can keep living our lives, creating and working, while reducing the risk posed by people who want to attack your values and the people you love.

---

[3] Only available in French for now (*Cyberattaques. Les dessous d'une menace mondiale*, Hachette, 2022).

[4] UK government, "UK Cybercrime Cost Report", https://www.gov.uk, 2011.

[5] World Economic Forum (WEF), "Global Risk Report", https://www.weforum.org, January 2023.

# THE BOOM MOMENT

In crisis management, the "Boom Moment" represents the precise point when a crisis materialises – the instant when a cyber attack, operational disruption, or reputational threat breaches preventive measures and demands immediate response. Originating from military terminology, where "boom" marks the detonation or impact of a threat, this concept has been widely adopted in corporate crisis preparedness to symbolise the critical shift from planning to active crisis response. The Boom Moment is when all contingency plans are put to the test, and the focus turns to limiting damage, restoring operations, and safeguarding assets.

**Left of Boom** encompasses everything leading up to that critical moment. Derived from counter-insurgency tactics, Left of Boom involves proactive measures to detect, prepare for, and ideally prevent a crisis. In cyber crisis preparedness, this phase includes risk assessments, employee training, security

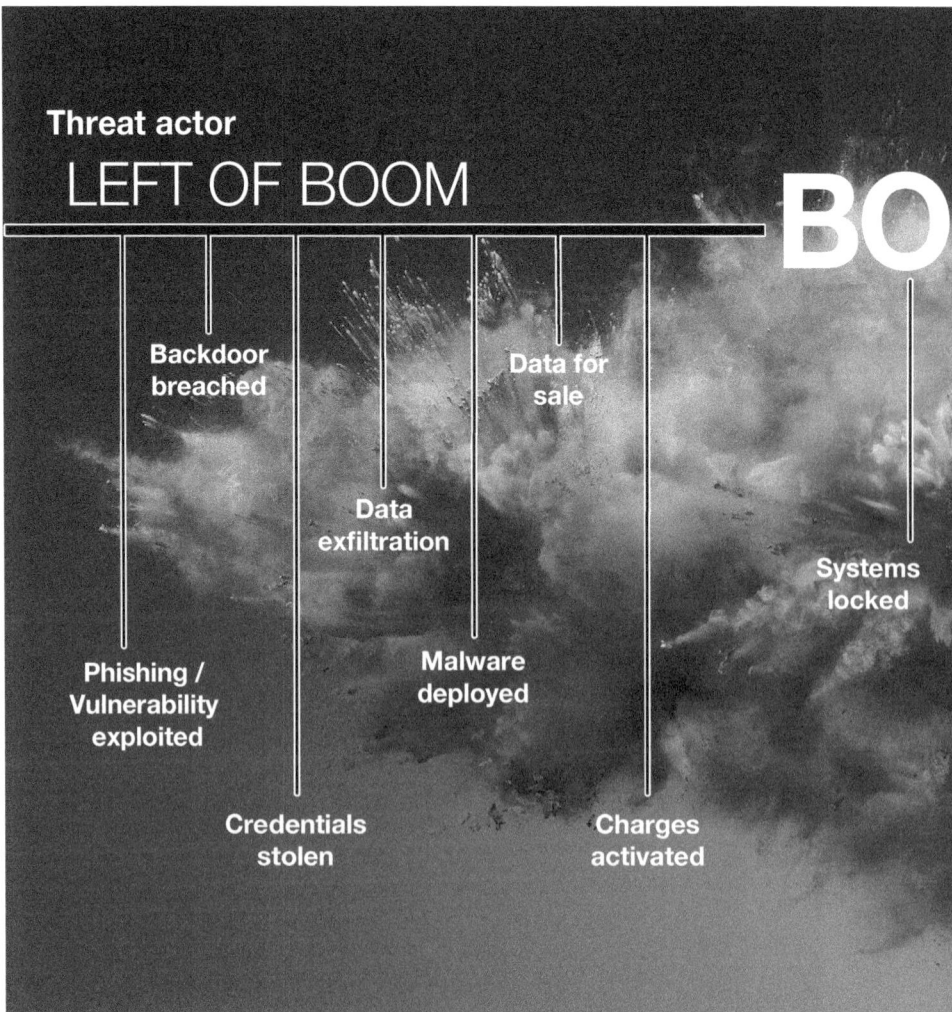

Threat actor

LEFT OF BOOM

BO

Backdoor breached

Data for sale

Data exfiltration

Systems locked

Phishing / Vulnerability exploited

Malware deployed

Credentials stolen

Charges activated

drills, and the development of robust detection and defence mechanisms. By staying Left of Boom, organisations focus on anticipation and prevention, aiming to identify and neutralise risks before they escalate.

However, if an incident does reach the Boom Moment, the focus shifts **Right of Boom** – the phase that covers response, containment, recovery, and post-crisis evaluation. Right of Boom in cyber crisis management is where all preparations are tested. This phase includes activating incident response plans, engaging communication protocols, containing the breach, and recovering affected systems. It's about minimising impact, restoring operations, and learning from the event to enhance future preparedness.

In essence, the Boom Moment divides the crisis life cycle into proactive (Left of Boom) and reactive (Right of Boom) phases. By building resilience Left of Boom, organisations can reduce the severity of Right-of-Boom actions, ultimately shortening recovery time and improving overall security posture.

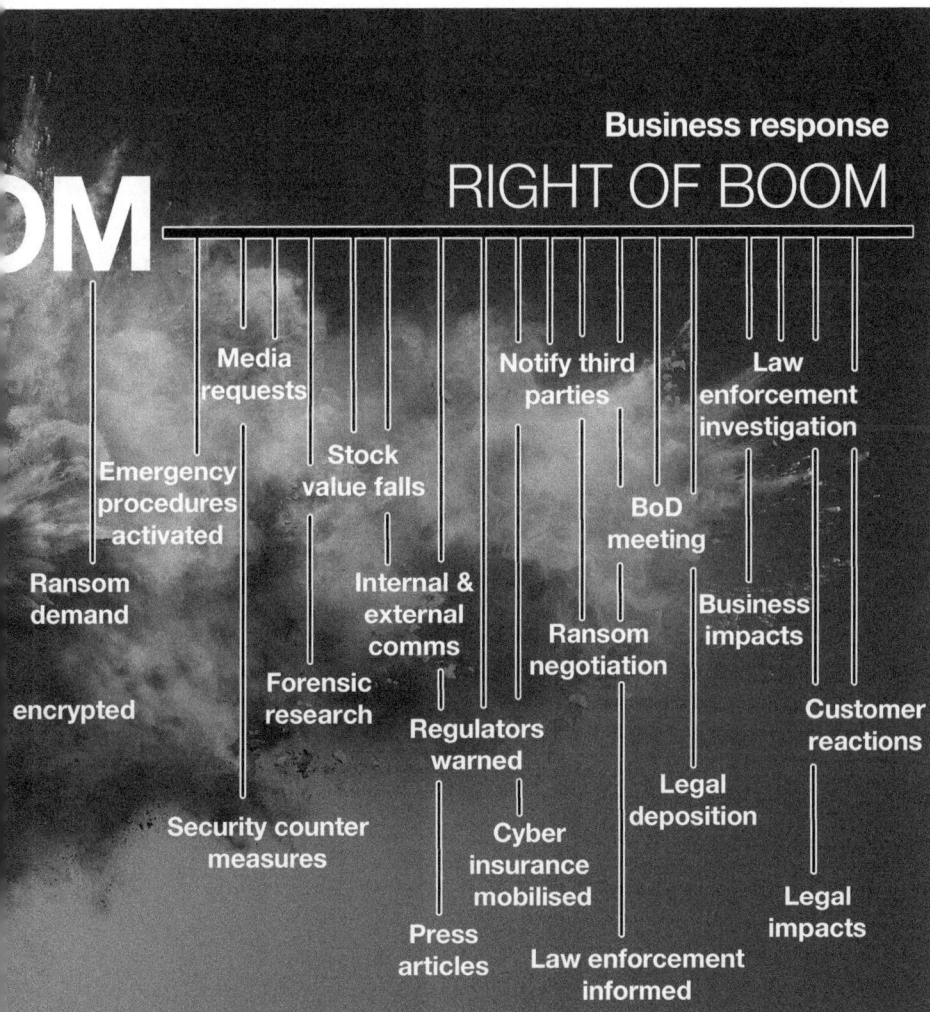

# 1

# THE DARK SIDE OF DIGITAL

"All human beings are capable of good and evil."

– Robert Louis Stephenson

For the past few years, we have been running a series of lectures titled "To Fight the Bad Guys, You Must Think Like the Bad Guys", at leading schools and universities in France and the UK. Our aim has always been to shift the mindsets of future business managers – whether in marketing, finance, or operations – regarding cyber security as a fundamental business issue. During one of these sessions, a student asked if there was an effective way to fully protect against cyber attacks. We smiled and answered, "Yes, there is one sure-fire way: you'd need a DeLorean time machine. Then put down your mobile phone and travel back to a time before computers existed." We added that while cyber crime is a modern phenomenon, it's still crime, and cyber security, at its core, is still about protecting what matters most.

## Preparation, Preparation, Preparation

Military history offers valuable lessons for any cyber-defence strategy. Preparation is essential. However, it's also critical to understand that zero risk does not exist. As long as humans rely on interconnected digital systems, cyber security professionals will never be out of work. History teaches us that any form of innovation, no matter how beneficial or ground-breaking, can be turned into a tool for harm.

Look at the twentieth century: scientists who pioneered nuclear physics likely never envisioned their discoveries being used to obliterate entire cities. A simple tool like a screwdriver, designed for fixing things, can just as easily be used as a weapon in the wrong hands. So, while digital transformation brings immense opportunity and innovation, it also introduces risks that are inevitable and, in many cases, unknowable.

*"Security isn't just a box to check."*

As the CISO of a major international chemical company points out, project and change management must integrate security concerns right from the start. He's seen first-hand that a company's resilience often depends on IT. Citing Margaret Thatcher's legacy in standardising government services with IT Infrastructure Library (ITIL), he reminds us that in today's digital landscape, security isn't just a box to check. "Embedding security within each project and change management process," he notes, "is the only way to mitigate growing digital risks before they become crises." By ensuring that security isn't an afterthought but a foundational component, organisations can adapt more confidently to the inevitable threats of digital transformation.

## We Can Only See the Tip of the Iceberg

If this sounds exaggerated or alarmist, think again. The headlines barely scratch the surface. For every reported cyber attack, there are countless others that go unreported or undetected. Most companies prefer to remain silent, fearing that their reputation will be damaged or that stakeholders will lose trust in them. But silence doesn't mean safety, and no one is immune – large or small, public or private.

In 2023, for example, the UK's National Health Service (NHS) was hit by a massive ransomware attack that crippled various IT systems, disrupting patient care across several hospitals and health services. This attack, which followed an earlier, smaller breach in 2017, highlighted the vulnerability of critical national infrastructure to cyber attacks. Despite extensive defences, there is no such thing as perfect security. Whether it's healthcare systems, financial institutions or small businesses, every sector is at risk.

Consider TalkTalk, a British telecom company that was hacked in 2015. Over 150,000 customers had their personal data exposed, leading to severe financial and reputational damage. The company was fined £400,000 by the Information Commissioner's Office (ICO) for its lax security measures, but the actual cost of lost business and consumer trust far exceeded that. And yet, TalkTalk is not an isolated case. British Airways suffered a major breach in 2018, which compromised the data of 400,000 customers and resulted in a record fine of £183 million under the General Data Protection Regulation (GDPR).

The common thread in these incidents? It's not a question of "if," but "when" the next attack will occur. No matter how prepared an organisation might be, the reality of today's digital world is that attacks are inevitable. The impact can be catastrophic: crippling business continuity, tarnishing reputations, driving down share prices, and in some extreme cases, leading to loss of life.

Do we sound far-fetched? Overly anxious? Perhaps, but the threat is real, and the evidence is everywhere. If you need proof, simply glance at the news – each week seems to bring fresh reports of yet another high-profile cyber attack. What we see in these headlines is just the tip of the cyber-crime iceberg. The reality is that many companies prefer to keep attacks quiet, fearing reputational damage and concerns over their ability to safeguard sensitive data.

No organisation is immune either, whether large or small, public or private. Well-known companies like British Airways, TalkTalk, Tesco, and WannaCry ransomware victims like the NHS have all been affected. Each of these incidents highlights that no matter how robust the cyber security protocols, it's inevitable that a breach will occur. When it does, the repercussions are broad and severe – business continuity is disrupted, reputations suffer, share prices drop, and even customer safety can be compromised.

In 2021, a cyber attack on Düsseldorf University Hospital in Germany tragically led to the death of a patient after the hospital's systems were compromised, preventing timely care. While the attack was not directly targeted at the hospital, the tragic consequences underscored the very real, human cost of cyber crime.

Zero risk is a myth, and as our world becomes more connected, predicting future threats becomes even more difficult.

With this harsh reality, more managers are seeking training in crisis management, eager to learn how to mitigate potential damage to their business. The reputational cost of making headlines as the victim of a cyber attack is immense, and no one wants to face that scrutiny. Nor do they want to explain to their customers, employees, or shareholders why their growth and investment plans have fallen short due to a preventable cyber incident. The personal consequences can be severe, too. In some jurisdictions, such as certain states in the US, company directors can be held criminally liable for data breaches that harm their customers. In extreme cases, CISOs and other cyber security leaders have been held personally accountable for breaches within their companies.

Could this legal trend spread to the UK? Potentially. With increasing regulatory scrutiny through bodies such as the ICO (Information Commissioner's Office) and laws like the GDPR (General Data Protection Regulation), the stakes for companies are higher than ever. Fines, like the £20 million penalty issued to British Airways following their data breach, serve as a sobering reminder of the financial and legal risks. It's not just about the IT department anymore – cyber security is now a board-level issue.

To drive this point home, we often pose a question in our training sessions: if your company experiences a cyber attack, do you think customers will blame the IT department or the brand itself? While it might seem rhetorical, the answer is clear. Customers rarely distinguish between internal departments – they hold the brand accountable. The impact on trust can be long-lasting, and rebuilding it requires considerable time, effort, and investment.

# Cyber Security Is a Business Issue

More and more business managers are seeking training in cyber crisis management, understanding that it's not just about preventing attacks, but mitigating their effects on the business. After all, no one wants their company to be the next NHS, British Airways, or TalkTalk – struggling to explain to shareholders, customers, and employees why projections are not being met and why recruitment and investment plans have been derailed. Worse still, some cyber incidents are personal. In several cases, Chief Information Security Officers (CISOs) and other senior managers have been held personally responsible for data breaches, a trend that could easily spread as accountability frameworks continue to evolve.

In the United States, some company directors have been held criminally liable for the damage caused by cyber breaches. The UK's ICO has also shown its willingness to enforce strict penalties under GDPR, with growing pressure on CISOs to ensure that their businesses adhere to the highest possible security standards.

## Who's to Blame?

When a cyber attack occurs, where does the blame lie? Is it with the IT department for failing to patch a vulnerability? Or does the responsibility fall on the brand itself? These are rhetorical questions, but ones that business leaders are increasingly forced to confront. Consumers rarely distinguish between technical failures and corporate responsibility. A reputation is built on trust, and once broken, it can take years to repair.

Marketing experts understand that acquiring a new customer is ten times more expensive than retaining an existing one. Imagine, for instance, your grandmother's connected pacemaker being hacked. You receive a message demanding payment in Bitcoin to stop the pacemaker from malfunctioning. The very idea is terrifying, but it's not as far-fetched as it might seem. Medical devices, critical infrastructure, and consumer products are all part of an expanding Internet of Things (IoT) network, creating new entry points for cyber criminals.

While ransomware (see Pocket Guide 1) has been widely discouraged by law-enforcement agencies, statistics suggest that businesses are often left with little choice but to pay, particularly when human lives or critical systems are at risk. As we'll explore in Chapter 2, paying a ransom once often means you'll be targeted again, creating a vicious cycle that can be hard to break.

## In the Blink of an Eye

The age-old saying that trust takes years to build and seconds to break has never been more relevant. In fact, companies like Apple have used their commitment to security as a central marketing theme. You may recall the shift in Apple's branding several years ago, when the company used a padlock to replace the iconic bite mark in its logo for certain ads, emphasising privacy and security. More recently, Samsung launched a mainstream TV advertising campaign focused heavily on mobile-phone data protection. These campaigns underscore the fact that cyber security is not just a technical issue – it's a brand issue. Data breaches may expose personal information, but the larger implications often include significant operational, financial, and legal consequences. In 2022, Optus, one of Australia's largest telecom providers, faced a major backlash after 10 million customer records were stolen. The breach not only attracted a public outcry but also resulted in the government moving forward with new legislation aimed at holding companies more accountable for protecting user data.

In the UK, Morrisons Supermarket experienced a similar crisis when a disgruntled employee leaked the payroll data of 100,000 staff members. While the ICO did not fine Morrisons, the case went all the way to the UK Supreme Court, which ruled that the company was not vicariously liable for the breach. Still, the reputational damage lingered, highlighting the importance of trust in an increasingly digitised world.

## Grasping the Nettle

As we explore throughout this book, digital transformation is not something that can – or should – be stopped. It's changing the way we live, work, and interact with the world. Every sector, from finance to agriculture, and from tourism to education, is undergoing a digital revolution, promising significant benefits. However, with that progress comes heightened risk.

To put things in perspective, SMEs (Small and Medium Enterprises) in the UK make up 99.9% of all businesses, and according to some insurers, the risk of going bankrupt increases by 50% in the six months following a cyber attack. A 2023 report by leading insurance broker Marsh[1] found that 45% of UK businesses identified cyber attacks as one of their top three risks, up from just 33% the year before.

The examples and statistics above demonstrate that cyber security is more than a technical challenge, it's a business imperative. Companies that fail to adapt to the evolving digital landscape and adequately protect themselves will struggle to survive in a world where connectivity comes with unprecedented risks.

## Progress and Risk Go Hand-in-Hand

In the UK, digital transformation has become central to strategies across various industries, from finance to manufacturing, healthcare, and even public services. The government itself launched a Digital Strategy aimed at transforming public services and enhancing the digital economy. The UK Digital Economy Council has also prioritised cyber security as a critical enabler of digital growth, emphasising the importance of securing the nation's digital infrastructure against the growing threat of cyber crime.

Recent Examples of UK Digital Initiatives

■ The NHS Digital Transformation: The UK's National Health Service (NHS) has been undergoing one of the most significant digital transformations in Europe. With a focus on digitising patient records, integrating artificial intelligence (AI) into diagnostics, and using big data to improve patient outcomes, the NHS has become increasingly dependent on digital technologies. However, as mentioned earlier, this reliance makes the NHS particularly vulnerable to cyber attacks. The 2017 WannaCry attack paralysed much of the NHS, costing the health service £92 million and highlighting the critical need for better cyber security measures as digital transformation continues.

■ Retail and E-commerce: During the COVID-19 pandemic, the UK saw an unprecedented shift towards online retail, with e-commerce sales reaching 30.2% of total retail sales in 2020, compared to 19.2% in 2019. Companies like Tesco and Sainsbury's quickly ramped up their online offerings, including mobile apps and home delivery services, but with this growth came new security concerns. In 2022, Tesco reported that hackers had attempted to steal customer details through a data scraping technique, leading to thousands of accounts being compromised.

---

[1] "Top risks for UK businesses revealed", https://www.marsh.com, 1 November 2023.

Although the company acted quickly to mitigate the damage, the incident served as a reminder of the security risks that come with rapid digital growth.

- The UK Public Sector: The GOV.UK Verify system, which allows citizens to securely access government services online, is another example of digital innovation in the public sector. However, like any digital service, it's vulnerable to attacks. In 2021, the NCSC (National Cyber Security Centre) warned that public sector organisations were being increasingly targeted by cyber criminals, particularly during the pandemic. Securing public services, which are now more reliant on digital platforms than ever, remains a top priority for the UK government.

# New Markets, New Risks

One of the key advantages of digital transformation is that it creates new markets and opportunities for growth. For instance, cloud computing and AI are unlocking new levels of efficiency and innovation. However, these very technologies are also expanding the attack surface for cyber criminals. Let's take a closer look.

- Cloud Computing: The rise of cloud computing has been a game changer for businesses across all sectors, enabling them to scale quickly, store massive amounts of data, and access software on-demand without investing in costly infrastructure. According to Statista, the global public cloud services market is expected to exceed $500 billion by 2025, with the UK being one of the largest cloud markets in Europe. However, this shift comes with risks, especially around data breaches and misconfigurations of cloud environments. In 2020, Marriott International was fined £18.4 million under GDPR after its cloud-hosted reservation system was breached, affecting 339 million customers. The breach was the result of weak security controls on its cloud infrastructure, which left sensitive data exposed for years. This serves as a cautionary tale for businesses rushing to adopt cloud solutions without properly securing their systems.

- Artificial Intelligence (AI): AI promises to revolutionise industries from healthcare to logistics, offering powerful tools for predictive analytics, customer service, and even cyber security itself. But AI also presents new challenges. AI-powered attacks, in which hackers use machine-learning algorithms to make their attacks more efficient and harder to detect, are on the rise. In the UK, Tesco Bank fell victim to an AI-enhanced cyber attack in 2016, in which hackers used sophisticated software to bypass the bank's fraud-detection systems, stealing £2.5 million from 9,000 customers. While AI can be a powerful tool for defending against cyber threats, it can also be used by attackers to automate and scale their operations, making it more important than ever for businesses to stay one step ahead.

## When Everything Goes Digital, the Risks Multiply

As we demonstrated earlier, glancing around we can see we are surrounded by digital and connected devices, both in our personal and professional lives. From smartphones and smart televisions to autonomous vehicles and medical devices, everything is connected; and that connectivity comes with a cost.

These everyday examples can be found on a much larger scale in every country and industry: agriculture, tourism, commerce, finance, education, health, manufacturing, defence, administration, and so on. Digital technology is simply everywhere.

One of the most concerning trends is the rise of Internet of Things (IoT) devices. By 2030, it's estimated that there will be over 125 billion connected IoT devices worldwide, each of which represents a potential vulnerability. Now I am not a mathematician but even I can tell you that's a large number. . . In the UK, smart meters have become a ubiquitous part of everyday life, with over 28 million installed across homes and businesses. While these devices help consumers monitor and reduce energy consumption, they also present a new avenue for hackers to disrupt critical infrastructure. In 2022, British Gas reported an attempted cyber attack targeting its smart meter system, which could have left millions of homes without power if it had been successful.

Connected devices are not limited to homes. Smart cities, equipped with connected infrastructure like traffic lights, CCTV cameras, and public Wi-Fi networks, are increasingly being targeted. The NCSC has warned that cyber criminals could exploit vulnerabilities in smart city systems to cause widespread chaos, from disrupting public transportation to cutting off utilities.

## The Threat to Critical Infrastructure

The impact of cyber attacks on critical infrastructure is becoming more evident with each passing year. In the UK, the National Grid, which controls the country's electricity and gas supply, has been a target of numerous cyber attacks, many of which are believed to have been state-sponsored. In 2021, SolarWinds, a software provider whose clients include government agencies and utilities, was breached in a sophisticated supply-chain attack. While the

UK was not as heavily affected as the United States, the incident underscored the vulnerabilities inherent in critical infrastructure.

The UK government has responded by tightening regulations, including the NIS (Network and Information Systems) Regulations 2018, which set out the requirements for protecting critical infrastructure. But as digital innovation accelerates, so too does the complexity of protecting these systems.

## Learning to Live with Digital Risk

Digital transformation is not just a technological shift, it's a cultural one. From retail to healthcare, education to manufacturing, every industry is being reshaped by digital innovation. But as we've explored in this chapter, with innovation comes risk. The more connected we become, the more vulnerable we are to cyber attacks. In the UK alone, the estimated annual cost of cyber crime exceeds £27 billion, with SMEs, public services, and critical infrastructure all in the crosshairs. While the benefits of digital transformation are clear – improved efficiency, lower costs, and new business models – it's critical that we acknowledge and prepare for the risks.

The adage holds true: it's not a question of "if" a cyber attack will happen, but "when". And when it does, businesses need to be ready to respond, mitigate the damage, and continue operating. In the next chapter, we'll explore the business model of cyber crime – understanding the motivations of attackers and how businesses can defend themselves in this high-stakes game.

### *Tourism*

In hotel management, digital transformation is reshaping how businesses operate, with a clear focus on enhancing customer experience, freeing up staff for more personalised services, and challenging traditional industry practices. Leading the way are online travel agencies such as Expedia and Booking.com, as well as community-driven platforms like Airbnb and Vrbo (previously known as Abritel), which have revolutionised how travellers book their accommodation.

Many global hotel chains are embracing fully connected, contactless technology, designed to enhance the guest experience while increasing operational efficiency. These advancements allow hotels to gather valuable insights about their customers, enabling a seamless and personalised stay. For instance, chains like Hilton and Marriott offer guests the ability to check in, access their rooms, and control in-room settings all via their smartphones. Premier Inn in the UK, for example, has introduced self-check-in kiosks, which reduces wait times and enables staff to focus on enhancing guest interactions.

At the forefront of this transformation are hotels that integrate high-tech amenities to streamline operations and improve guest comfort. Take Pullman Hotels, where guests can use touch-sensitive tablets to control room temperature, lighting, and even communicate with hotel services. Similarly, Accor and Hyatt hotels now allow guests to check in 100% digitally and use their smartphones as room keys, eliminating the need for traditional key cards and front-desk interactions.

This digital shift is not just limited to global hotel chains; smaller, boutique hotels are also adopting these technologies to stay competitive. In the UK, The Hoxton and Z Hotel Group have incorporated digital tools to enhance guest experiences, from smart room controls to tailored recommendations via mobile apps.

The international hospitality sector recognises the significance of this transformation. As Les Roches, a world-renowned hotel management school, states, "Transformation means the accelerated adoption not of futuristic technologies, but of existing technological applications such as automation, artificial intelligence, the Internet of Things, robotics, and blockchain." The Boston School of Hospitality Administration adds, "We are currently living through the fourth industrial revolution, a period marked by emerging technologies such as artificial intelligence, robotics, virtual reality, the Internet of Things, and fifth-generation wireless technology. This period has dramatically transformed our hospitality industry and will continue to do so."

In the UK, digital innovation in the tourism and hospitality sector has been further accelerated by the COVID-19 pandemic, prompting many businesses to adopt contactless solutions as a matter of health and safety. Travelodge and Holiday Inn were among the first to introduce mobile check-in and digital key options in response to new hygiene standards. Digital tools have also enabled tourism agencies and destinations to offer virtual tours, interactive travel guides, and more personalised online services, providing a glimpse into the future of travel and hospitality in a post-pandemic world.

## Trade

Trade, particularly in e-commerce, has undergone a profound transformation over the past decade. According to INSEE (the French National Institute of Statistics and Economic Studies), e-commerce encompasses "all sales transactions using the internet or other IT networks such as the exchange of computerised data and involving a change of ownership of the good or service being ordered. Goods and services are ordered via these networks, but payment and final delivery of the good or service may be carried out using traditional methods."

E-commerce has become an integral part of daily life, driven by consumer demand for convenience and accessibility. Major online retailers such as Amazon and eBay have seen substantial growth over the years, and the rise of UK-based digital retailers such as Ocado has underscored the increasing importance of logistics and digital platforms in sustaining trade. Ocado's advanced use of robotics and AI in warehousing and delivery showcases the impact of digital transformation on trade efficiency.

In the UK, the e-commerce sector continues to thrive, with Royal Mail reporting sustained increases in parcel volumes as online shopping becomes the norm for many consumers. Meanwhile, smaller businesses, particularly those that traditionally relied on physical retail, have successfully transitioned to digital platforms such as Etsy and Notonthehighstreet, allowing them to reach wider audiences and diversify revenue streams.

This shift has led to more sophisticated web portals, saved payment methods, and a seamless, multi-channel purchasing experience for customers. Retailers such as ASOS and John Lewis have embraced these trends by developing mobile-friendly websites, introducing fast-checkout options, and offering flexible delivery services. Meanwhile, companies like Deliveroo and Just Eat have expanded their services to include not only food but also everyday essentials, further integrating digital solutions into traditional shopping habits.

In its recent report on digital trade,[2] the OECD (Organisation for Economic Cooperation and Development) confirms that digital transformation is increasing the scale, scope, and speed of global commerce. It enables businesses to introduce new products and services to a growing number of connected consumers, both domestically and internationally. This trend has been particularly impactful for UK businesses, many of which have turned to digital tools to overcome traditional trade barriers, streamline payments, and foster international collaborations. Small and medium-sized enterprises (SMEs) have benefited from cloud services and crowdfunding platforms that provide them with the flexibility and capital to innovate without significant upfront investments in infrastructure.

---

[2] "Of bytes and trade: Quantifying the impact of digitalisation on trade", OECD Trade Policy Paper, www.oecd.org, no 273, May 2023.

At the same time, digital technologies and innovative business models are reshaping the production and delivery of services. The rise of subscription-based models, such as those employed by companies like Spotify and Netflix, demonstrates how services are increasingly crossing the traditional lines between products and services. ASOS Premier, for instance, offers unlimited next-day delivery for a fixed annual fee, blurring the line between logistics and retail offerings.

This digital shift is not without its challenges. The rapid growth of e-commerce has highlighted the need for more efficient last-mile delivery solutions, where companies like DPD and Hermes are investing in electric vehicles and AI-driven route optimisation to meet demand while maintaining environmental sustainability. Concerns around data privacy and cyber security are growing, as retailers collect and process ever more customer data to enhance personalisation and streamline their supply chains.

In essence, digital transformation is not only reshaping how goods and services are produced and delivered, but it is also redefining the very nature of trade itself. As the *Financial Times* reported in 2022, digital trade agreements, like the UK-Singapore Digital Economy Agreement, are paving the way for smoother cross-border trade and enhanced digital cooperation, marking a new era in global commerce.[3]

## Finance

What would modern finance be without digital technology? Technology has become the backbone of banking, insurance, and broader financial operations. From multi-channel customer interactions and online banking to direct banking, payment apps, and industrial-scale cheque processing, every aspect of financial services is deeply intertwined with digitisation. The sector's reliance on digital technology is so entrenched that even behind-the-scenes processes, like data centre operations, are powered by cutting-edge technology, robotics, and automated systems.

A Microsoft report published in December 2022 took a five-year perspective on the digital transformation of the financial sector. The statistics tell a compelling story:

■ 80% of companies believe that understanding their customers is a key external factor in driving digital transformation – an increase of 7% compared to 2017.

---

[3] "UK and Singapore sign new innovative digital trade deal", 25 February 2022, https://www.gov.uk/government

- 80% of respondents also view being seen as innovative as a critical factor in accelerating digital transformation – an 8% rise since 2017.

- 74% of companies state that digital transformation is a means of attracting and retaining talent – up 17% from five years ago.

- 62% of CIOs (Chief Information Officers) are now involved in shaping their organisation's digital transformation strategy, a significant rise from 24% five years prior.

- 35% of companies feel that there IS (information system) are well-suited to digital transformation projects – an increase of 18%.

- 71% of companies and SMEs are willing to trust sovereign cloud solutions, even if these use technology from large, international providers, including American firms.

These statistics underscore the clear importance of digital transformation in finance. But the journey isn't always smooth. Many finance departments are still grappling with the need to change long-standing habits and workflows to accommodate new technologies.

A Gartner report titled "The State of Digital Transformation for Financial Services Business-Line Leaders"[4] highlights the magnitude of this shift, reporting that 69% of business leaders indicate that digitisation programmes are on the rise, and many predict that digital technologies will radically transform their industry by 2026.

The financial sector is evolving rapidly, and with it, traditional approaches are being reshaped. Fintech innovations such as blockchain, artificial intelligence (AI), and machine learning are being integrated into everything from fraud detection to customer service automation, bringing new efficiencies and capabilities to the industry. UK-based challenger banks such as Monzo and Revolut have fully embraced AI-driven solutions to enhance customer experience and reduce operational costs.

In addition, financial institutions are increasingly turning to the cloud to power their operations. According to IBM, cloud computing offers finance companies substantial advantages, including:

- Reduced IT costs: The cloud allows companies to offload infrastructure costs, such as purchasing and maintaining servers, onto cloud service providers.

---

[4] "The state of digital transformation for financial services business-line leaders", www.gartner.com, 2019.

■ Increased agility: Cloud services allow financial institutions to rapidly deploy new applications, reducing time-to-market for innovative services.

■ Scalability: Cloud platforms enable financial institutions to scale resources up or down in response to demand, making operations more efficient.

However, along with these advantages, increasing digitisation brings a growing concern over cyber security and data privacy. The financial sector is a prime target for cyber attacks due to the vast amounts of sensitive customer data it processes daily. As financial institutions become more digitally advanced, the tactics of cyber criminals follow suit, prompting an ongoing arms race to secure data, systems, and digital assets.

As the UK's Financial Conduct Authority (FCA) highlighted in its 2023 report on cyber security in finance,[5] ensuring that institutions are not only prepared for digital transformation but also adequately protected against cyber threats is of paramount importance. A single breach can result in millions of pounds in losses, damage to reputation, and a potential loss of customer trust.

---

**DORA (no, not that one. . . )**

One of the most significant recent developments in the finance sector's digital transformation is the introduction of the Digital Operational Resilience Act (DORA)[6] by the European Union. Though this regulation applies primarily to EU firms, it has implications for UK-based finance firms that operate in or interact with EU markets, including those providing services or working with EU-based partners.

DORA was introduced to strengthen the financial sector's resilience to ICT (Information and Communication Technology) risks, including those posed by cyber security threats. It sets out to create a harmonised and comprehensive framework that ensures all entities within the financial services sector can withstand, respond to, and recover from ICT-related disruptions and incidents.

**Let's Explore DORA**

■ Governance and oversight: DORA requires firms to have governance frameworks in place to properly manage digital operational risks. This includes appointing senior management responsible for overseeing ICT risks and ensuring they are integrated into the firm's overall risk-management framework.

▶

---

[5] "2023 CBEST thematic", www.bankofengland.co.uk

[6] Regulation (EU) 2022/2554 of the European Parliament and of the Council of 14 December 2022 on digital operational resilience for the financial sector and amending Regulations (EC) No 1060/2009, (EU) No 648/2012, (EU) No 600/2014, (EU) No 909/2014, and (EU) 2016/1011, *Official Journal of the European Union*, https://eur-lex.europa.eu, 27 December 2022.

- Incident reporting: One of the core elements of DORA is the mandatory reporting of ICT-related incidents. Firms are required to categorise and report incidents in a timely and structured way to the relevant authorities, enhancing transparency and enabling a quicker, coordinated response to significant events. UK-based financial institutions with EU customers or counterparts must comply with these reporting obligations when relevant.
- Third-party risk management: DORA places special emphasis on the management of third-party risks, particularly in relation to critical ICT service providers. Financial entities are now required to ensure that third-party ICT providers, such as cloud services, meet strict operational resilience standards. This aligns with the broader trend of scrutiny on outsourced services within the UK's financial sector, including under guidelines from the Bank of England and the FCA (Financial Conduct Authority).
- Penalties for non-compliance: DORA introduces tough penalties for firms that fail to comply, including financial penalties and restrictions on operating activities. These consequences serve as a strong incentive for firms to bolster their digital resilience strategies.

**Impact on the UK Finance Sector**

Although DORA is an EU regulation, its impact on the UK finance sector is substantial. Many UK firms, particularly those with operations in Europe or dealing with EU customers, will need to comply with DORA's provisions. For UK-based firms that fall outside of direct EU jurisdiction, the principles embedded in DORA reflect a broader shift towards enhanced regulatory scrutiny around digital resilience.

UK regulators such as the FCA and PRA (Prudential Regulation Authority) have signalled a commitment to enforcing similar standards in the UK. While the UK has not adopted DORA, it shares many of the same goals, and firms are expected to meet high standards of resilience through existing UK regulations, such as the Operational Resilience Framework.

In practice, this means UK financial services firms are likely to face similar requirements, including stringent operational risk management and incident reporting rules. This is part of a broader global move toward ensuring that financial institutions are not only prepared for cyber attacks but can also maintain operational continuity if they do occur.

**DORA and Digital Transformation**

Digital transformation has been rapidly reshaping the finance industry, but it has also exposed firms to unprecedented risks. DORA provides a framework for managing these risks while allowing firms to innovate and continue their digital journeys securely. For example:

- Cloud services: Cloud computing offers major efficiencies, but DORA places significant emphasis on the oversight of cloud providers. Companies must

▶

ensure that cloud vendors meet the same rigorous security and resilience requirements as their internal systems.

■ Cyber security requirements: DORA's emphasis on cyber security and incident reporting aligns with the growing pressure on financial institutions to improve their digital defences. As UK firms increasingly rely on digital services, the ability to monitor, report, and mitigate cyber security incidents will be crucial for maintaining both regulatory compliance and consumer trust.

DORA represents a significant shift in how financial institutions approach operational resilience and digital transformation. While it is a EU regulation, its principles are increasingly shaping regulatory landscapes beyond the EU, including the UK. UK-based firms, especially those with cross-border operations, will need to ensure that their digital strategies are aligned with both DORA and the UK's evolving regulations around operational resilience.

As digital transformation continues to reshape the finance sector, robust frameworks like DORA will play an essential role in maintaining stability and trust, not just in Europe, but globally.

## Health

In the UK, digital transformation in healthcare has taken centre stage with the development of the NHS Digital service, which is responsible for implementing and overseeing the digital strategy of the NHS (National Health Service). According to NHS Digital, the mission is to "improve lives by building a better healthcare system through digital technology". This includes the integration of electronic health records (EHR), data analytics for patient care, and digital platforms that allow for remote consultations – which became particularly essential during the COVID-19 pandemic.

The UK government's NHS Long-Term Plan highlights a strong commitment to advancing digital health, with a specific focus on the use of artificial intelligence (AI) to assist in diagnosis and treatment, improving patient access to services through apps like NHS 111, and enhancing care by integrating patient data across the system. This initiative aligns with the global trend of digital transformation, where the e-health market is expected to grow exponentially. According to Tech Nation,[7] the UK's health tech sector grew by 61% between 2020 and 2021, showing how the UK's focus on digital health has accelerated.

---

[7] "Lifting the lid on how UK tech boomed in 2020", Tech Nation, 2021 Report, https://growth.technation.io, 23 November 2021.

Additionally, the UK has seen significant investment in digital health technology. A notable example is the £1 billion Digital Health and Care Plan, which was unveiled to boost the adoption of digital tools and platforms across the NHS. This includes investments in telemedicine, AI-based diagnostic tools, and the modernisation of healthcare infrastructure to improve both patient outcomes and operational efficiency.

On the private sector side, digital transformation is also reshaping healthcare delivery in the UK. For instance, private healthcare providers like Bupa and Spire Healthcare have introduced fully digital platforms for consultations, patient records, and virtual GP services. These changes have been driven by patient demand for more convenient, faster access to healthcare services. With companies increasingly utilising connected devices, wearable tech for health monitoring, and cloud-based systems to store patient data securely, the healthcare landscape in the UK is rapidly evolving.

The COVID-19 pandemic significantly accelerated this shift. Matt Hancock, former UK Health Secretary, had described the rapid deployment of telemedicine during the pandemic as a transformation that would have taken a decade without the crisis. Remote GP consultations, which surged during the pandemic, are expected to remain a standard part of UK healthcare. The government's Health Infrastructure Plan outlines a commitment to investing in digital tools and AI to support a more efficient and resilient health system.

The private sector has similarly recognised the need for digital health solutions to cope with crises and beyond. Sanofi UK, for example, leveraged data analytics and digital tools during the pandemic to help manage distribution networks and respond to shifts in patient demand for medicines and vaccines. The healthcare company also adapted its workforce to remote operations, demonstrating how digital health technologies can maintain continuity of care and support public health objectives during unprecedented times.

## Manufacturing

Digital technology is critical to the future of manufacturing in the UK. It has transformed working methods, organisational structures, and production processes across industries. The advent of connected factories (also known as smart factories), 3D printing, robotics, and extended reality has revolutionised production cycles, enabling companies to remain competitive in the global market. In the UK, initiatives like Made Smarter, a national programme, help manufacturers adopt these cutting-edge technologies to improve productivity and reduce costs. These changes touch every part of the organisation, from design optimisation using data digitisation to production automation through robotics, as well as quality monitoring and predictive maintenance supported by Internet of Things (IoT) sensors and big data analytics.

An excellent example of digital transformation in UK manufacturing is Rolls-Royce, which uses digital twins to create virtual models of engines. These digital twins enable predictive analysis, helping the company to identify and address performance issues before they result in equipment failure, thus improving operational efficiency and reducing downtime. Similarly, BAE Systems is leveraging automation to streamline its production processes, particularly in the aerospace sector, where precision and safety are paramount.

Additive manufacturing (3D printing) has become a game-changer in the automotive and aerospace industries. For example, Jaguar Land Rover has incorporated this technology into its prototyping process, reducing waste and speeding up design iterations. Siemens, another leader in the manufacturing sector, has applied IoT to enable more predictive, efficient maintenance of machinery across its factories, part of its Factory of the Future programme.

And the High Value Manufacturing Catapult, an organisation dedicated to boosting UK manufacturing, has played a significant role in fostering innovation by supporting SMEs and large manufacturers alike in adopting advanced digital technologies such as AI and 5G. The Catapult's digitalisation initiatives have been instrumental in helping manufacturers reduce operational costs, improve sustainability, and increase productivity.

The UK government is strongly committed to digital transformation in manufacturing. The Industrial Strategy Challenge Fund supports investments in automation and AI across various industries, with a specific focus on sectors like automotive and steel. Furthermore, advanced robotics is helping manufacturers optimise complex tasks on the production floor, where precision and efficiency are critical. These technologies are also seen as key enablers in reducing carbon emissions and promoting resource efficiency, in line with the UK's ambition to achieve net-zero emissions by 2050.

Digital skills are crucial to this transformation. Universities and research institutes across the UK are offering specialised courses to train future engineers and digital specialists. For example, the University of Sheffield is known for its Advanced Manufacturing Research Centre (AMRC), which helps businesses adopt new technologies to stay competitive.

Overall, digital technology has radically changed the manufacturing landscape in the UK, offering both opportunities and challenges. As more companies integrate these advancements into their operations, the future of UK manufacturing looks increasingly high-tech, efficient, and globally competitive.

## *Defence*

In the UK, digital transformation in the defence sector has been a top priority, and it plays a crucial role in modernising military capabilities. The Ministry of Defence (MoD) is embracing digital technology to ensure that the UK's armed forces maintain a technological edge. Digital transformation initiatives like Defence Digital, led by the MoD, focus on enhancing data management, cyber resilience, and AI integration within military operations.

One key area of innovation is the deployment of AI in defence systems, which helps to optimise decision-making in real-time, ranging from strategic operations to autonomous defence platforms. For instance, Project DART within the MoD has focused on incorporating AI to provide better intelligence support in combat scenarios. AI and big data analytics are being used to identify patterns in surveillance data, predict threats, and enhance operational efficiency. This is part of the Integrated Operating Concept, which aims to adapt the UK's military structure to the modern era of multi-domain operations.

Another important initiative is the UK's work on autonomous drones. For instance, the Tempest programme, led by BAE Systems, Rolls-Royce, Leonardo, and MBDA, focuses on developing the next-generation fighter jets equipped with AI capabilities and autonomous drone swarms. This aligns with the Future Combat Air System (FCAS) strategy, which aims to deliver technologically advanced, AI-driven aircraft by 2035.

The MoD's Digital Strategy for Defence, published in 2021, also emphasises cyber resilience, outlining plans for Zero Trust Architecture – a concept where no user or system is trusted by default when interacting with defence networks. This strategy aims to prevent data breaches and ensure the security of sensitive military information, as cyber attacks become increasingly sophisticated.

The use of data analytics and cloud computing is further enhancing the operational capabilities of the UK's armed forces. The Strategic Command, a component of the MoD, is actively pursuing cloud technologies to ensure seamless access to data across various military operations. These capabilities are critical for military effectiveness, allowing for quicker decision-making and more agile responses to emerging threats.

The role of robotics in defence has also expanded. Semi-autonomous robots are now being used for mine clearance, logistics, and even medical evacuation, reducing the risk to human personnel in hazardous environments. The Defence Science and Technology Laboratory (DSTL) is actively researching and deploying robotic technologies, which provide operational advantages on the battlefield.

In line with these innovations, the UK's Defence Cyber Security Strategy ensures that digital advancements are paired with robust cyber security measures. The MoD is heavily investing in cyber security skills, training personnel to defend against growing cyber threats, including state-sponsored attacks and cyber terrorism. Cyber resilience has been recognised as a key component of national security strategy, with GCHQ and Cyber Command playing central roles in safeguarding the UK's digital defence infrastructure.

# But. . . Is Innovation Always Good for You?

The march of digital transformation often leads to a fundamental question: Is all innovation truly beneficial, or does it sometimes pose new risks that outweigh its advantages? As we've discussed, technological innovations bring with them undeniable improvements, but they can also be used for harm when in the wrong hands.

## *The Food Industry*

One of the latest trends in the agri-food market is personalisation. Consumers are increasingly expecting custom nutrition solutions based on detailed dietary advice and data-driven platforms. According to a report by Global Data,[8] 64% of consumers around the world have discovered ways to personalise their shopping experience to align with their beliefs and individual needs. In the UK, major companies like Unilever and Nestlé are leading initiatives that use consumer data to tailor product offerings. For example, Nestlé launched its personalised nutrition platform, Nestlé Wellness Club, which provides individualised dietary advice and product recommendations based on consumer health data.

However, these personalisation efforts rely heavily on data collection, raising concerns about privacy and security. The more the data collected, the greater the responsibility to protect it (like that line from Spider-Man, "With great power comes great responsibility"). Under the General Data Protection Regulation (GDPR), companies handling consumer data in the UK are required to safeguard it against misuse or unauthorised access. Failure to do so can result in significant reputational damage and hefty financial penalties. For example, British Airways was fined £20 million in 2020 for failing to protect customer data under GDPR.

If this data falls into the wrong hands, it could result in reputational damage and potential financial penalties under GDPR.

---

[8] "Nearly half of global consumers are influenced by changes in society when purchasing products, says GlobalData", www.globaldata.com, 24 March 2022.

## Telecommunications

One of the prevailing trends in the telecommunications market is the surge in content creation and distribution. The global market for content distribution networks is projected to double, increasing from $15 billion to $30 billion by 2025,[9] driven in part by streaming consumers – both businesses and individuals – who are central to this growth.

In the UK, major players such as BT and Sky are heavily investing in Content Delivery Networks (CDNs) to enhance streaming capabilities and meet the growing demand for high-quality video and media services. BT Sport, for example, has expanded its streaming services to deliver 4K Ultra HD content to its users. As demand for streaming services accelerates, the need for faster, more efficient content delivery becomes essential.

This rapid expansion, however, also introduces new vulnerabilities. The collection, analysis, and use of ever-larger datasets raise concerns around data privacy and security. Telecommunications companies are at the forefront of gathering vast amounts of user data to optimise service delivery. As a result, cyber criminals are increasingly targeting these networks, exploiting potential weaknesses to gain access to sensitive consumer information or disrupt services. In fact, the UK's National Cyber Security Centre (NCSC) has warned that telecoms infrastructure remains a high-value target for cyber attacks, particularly given the sector's role in facilitating national connectivity.

## Automotive

Autonomous vehicles (AVs) are certainly causing a stir and presenting substantial investment opportunities within the automotive sector. The allure of self-driving freedom is undeniable, and AVs are poised to reshape the relationship between drivers and their vehicles. These vehicles are gradually being introduced to the market with varying levels of automation, which allows time for societal adjustment, regulatory developments, and shifting attitudes toward autonomous driving.

While optimism for AVs is strong, some setbacks remind us of the risks associated with the technology. For instance, a tragic accident occurred in 2018 involving an Uber-branded autonomous vehicle in Arizona, USA, which

---

[9] "Content Delivery Network Market Size & Share Analysis – Growth Trends & Forecasts Analysis (2024-2029)", Mordor Intelligence, https://www.mordorintelligence.com, 2023.

resulted in a pedestrian fatality. Since then, advancements have been made, but challenges remain. In late 2022, the US National Highway Traffic Safety Administration (NHTSA) began investigating incidents involving General Motors' autonomous taxis, which had reportedly experienced sudden braking issues.

In the UK, autonomous vehicle testing has accelerated. Projects such as the UK Autodrive programme have successfully trialled AVs in real-world settings, with participants like Jaguar Land Rover and Ford taking part in trials in Milton Keynes and Coventry. These trials aim to assess the technology's safety and its ability to navigate UK roads under real-world conditions. The government's plans to roll out AV technology have the potential to boost the automotive sector's competitiveness, as outlined in the UK's Industrial Strategy.

Meanwhile, reports of autonomous vehicle incidents are not uncommon. Tesla, for example, was the subject of an investigation in 2022 after its models equipped with Level 2 autonomous driving technology were involved in over 270 accidents. Level 2 systems allow the car to manage basic tasks such as steering and braking but still require human oversight. As such, human error continues to play a significant role in AV safety, even as automation advances.

As the UK government moves forward with its ambition to have self-driving cars on the roads by 2025, the regulatory framework and safety considerations will remain critical. Ongoing collaboration between automakers, regulators, and tech firms will be necessary to ensure that AVs can safely integrate into everyday life.

## Manufacturing

Many UK manufacturers are pushing for intrusion detection systems, typically found in IT platforms, to be extended to factories. This is because the manufacturing industry is increasingly adopting the Internet of Things (IoT), connecting previously isolated devices to wider digital networks. The IoT can improve efficiency, reduce costs, enhance security, and ensure compliance, all while driving product innovation. However, this increased connectivity also brings significant challenges. Industrial systems generally have much longer upgrade cycles than IT systems. Often, software updates – frequently labelled as "safety improvements" – are provided solely by the Original Equipment Manufacturer (OEM). The fear is that by not applying the latest updates, manufacturers risk losing warranty support. Additionally, these industrial tools typically have long life spans, sometimes over fifteen years, leaving them vulnerable to obsolescence if security weaknesses emerge. This creates a challenging environment where manufacturers must balance operational stability with cyber security.

Another concern is the rapid deployment of IoT devices and their integration into manufacturing processes adds complexity. The speed of IoT development often outpaces traditional industrial safety protocols, making it harder to ensure that every connected device meets stringent security standards. This opens up vulnerabilities that, if exploited, could lead to significant production disruptions or safety concerns. As such, while the benefits in manufacturing are clear, the rapid pace of technological innovation presents new risks, requiring manufacturers to remain vigilant.

## Medicine

*Forbes* published an article in 2020[10] that explored an emerging concept: precision medicine. While we always knew that medicine was a precise discipline, the idea that drugs could one day be personalised to match each patient's genetic profile was ground-breaking. This approach has the potential to make treatments more effective and reduce the likelihood of unwanted side effects. However, scaling this concept into widespread practice presents significant challenges, particularly in terms of data management.

The personalised nature of precision medicine demands the collection, analysis, and use of vast amounts of sensitive data, including genetic information, with patients' informed consent. Pharmaceutical companies and healthcare providers will need to safeguard this data rigorously, adhering to stringent privacy regulations like the UK's General Data Protection Regulation (GDPR).

What could possibly go wrong. . . ? The risks are evident – any compromise of such data could result in severe privacy violations, financial penalties, and loss of public trust.

## Public Sector

The management of drinking water is a critical issue for current and future generations. In developed countries, public expectations for water quality are high, often surpassing concerns about air quality. Even a minor odour, taste, or slight discolouration can trigger suspicion about the safety of the water supply.

---

[10] Budel, S., "The precision medicine revolution will be driven by diagnostic technologies", *Forbes*, 28 April 2020.

This growing awareness has pushed governments to modernise ageing drinking-water infrastructures, many of which are decades old. To address these challenges, advanced, connected sensors are being deployed to provide real-time monitoring and ensure the safety and quality of water systems. These systems use the Internet of Things (IoT) to detect any contamination and address issues before they pose a significant health risk.

As water infrastructure becomes increasingly digitised, it also becomes vulnerable to cyber attacks. Two notable incidents illustrate the danger. In early 2021, hackers attempted to poison a water treatment plant in Oldsmar, Florida, by altering the chemical levels in the water supply. In mid-2022, a major British water supplier faced a cyber attack, raising concerns about the security of critical infrastructure in the UK. These events underscore the importance of cyber security in the modernisation of public utilities.

Another example of a critical issue in the modern era is energy management, highlighted by the call for greater control over energy consumption made by EU Member States at the end of 2022. To achieve this goal, the economic and regulatory framework will need to adapt, and energy networks will have to become increasingly intelligent. Smart meters, already a key feature in many households, offer more precise control over energy consumption and generation, tailoring use to the needs of both suppliers and consumers.

To make electricity production and distribution more efficient on a global scale, vast amounts of data must be collected, analysed, and utilised. Smart grids, connected meters, and advanced analytics enable more precise predictions of energy demand and real-time adjustments. This approach also helps integrate renewable energy sources into the grid, reducing reliance on fossil fuels.

This push for smarter, interconnected electricity systems brings with it inherent challenges. Older infrastructure and interconnecting legacy systems will need to be upgraded or replaced to handle the complexities of data-driven energy management. These updates increase the operational efficiency of energy networks but also expose them to the risk of cyber attacks, making cyber security a crucial component of modern energy strategies.

## Fomo and Innovations to Come

We are surrounded by digital transformation – if nothing else, we've established that so far in this book. There's no need to debate this trend further, it's an empirical reality. We have embraced digitalisation for several decades now, and according to IT experts and with colossal investments in this sector, it is clear this trend will only accelerate.

## Could IT have Its head in the clouds?

Cloud computing is a prime example of digital transformation. You've probably come across terms like private cloud, public cloud, hybrid cloud, and sovereign cloud. These aren't abstract, distant concepts but concrete systems shaping the future of data management. Major players like Amazon, Google, IBM, and Microsoft dominate the cloud space, alongside national suppliers that cater to specific markets. Cloud computing has allowed businesses to migrate from on-premise infrastructure to digital environments hosted by Cloud Service Providers (CSPs), offering significant benefits such as reduced costs, scalability, and improved agility.

In the UK, cloud adoption is widespread across industries and public sectors. The UK government's G-Cloud framework enables public sector organisations to procure cloud services more easily, demonstrating the government's push toward digital transformation. According to IBM, cloud computing delivers IT resources, storage, and development tools via CSPs, allowing businesses to pay for services on a monthly or per-use basis, helping organisations optimise costs.

Cloud services provide three primary models, according to Google Cloud:

- IaaS (Infrastructure as a Service): The CSP manages the back-end infrastructure (computing, networking, storage), while the user manages the operating system and applications.
- PaaS (Platform as a Service): The CSP handles the back-end infrastructure and software tools needed to develop applications, with users still managing their own code and data.
- SaaS (Software as a Service): The CSP delivers the entire application stack, allowing users to connect directly without managing the back-end systems.

In the UK, cloud computing has been instrumental in driving down IT costs for businesses and public sector organisations alike. The National Cyber Security Centre (NCSC) provides a set of cloud security principles that encourage secure cloud adoption across government departments and private companies, helping mitigate potential vulnerabilities.

From a security perspective, the cloud is seen as a way of improving vulnerability management, as CSPs take responsibility for software patches and updates.

However, because cloud computing is still relatively new, configuration errors have led to numerous high-profile breaches. For instance:

■ Amazon: On June 17, 2020, a record-breaking distributed denial-of-service (DDoS) attack targeted Amazon's cloud service with 2.3 Tbps of traffic, marking the largest DDoS attack ever recorded.

■ X (formerly Twitter): On June 24, 2020, X disclosed that billing information was accidentally stored in browser caches, exposing sensitive data, including phone numbers and partial credit card details.

■ Pfizer: On July 9, 2020, a breach due to a misconfigured Google Cloud bucket exposed the medical records of over 100 patients.

■ Kaseya: On July 2, 2021, a ransomware attack crippled Kaseya's clients, leading to widespread outages and forcing many businesses to halt operations.

■ Uber: On September 16, 2022, Uber's cloud partner, Amazon, was compromised, leading to the leak of employee emails and internal reports.

These incidents illustrate that despite its advantages, the cloud's shared responsibility model leaves room for errors and misconfigurations. In the UK, the NCSC emphasises the importance of proper configuration and user responsibility when using cloud services.

Cloud adoption is forecasted to continue growing, especially with emerging trends such as sovereign cloud – a cloud solution designed for specific countries to address data sovereignty issues. With the UK's strong push towards digital transformation, organisations are increasingly turning to cloud services to maintain business continuity, enhance agility, and manage resources more efficiently.

Service Level Agreements (SLAs) define the relationship between CSPs and customers, outlining metrics like uptime, response times, and performance standards, which ensure accountability for both security and performance. Microsoft's shared responsibility model highlights that, depending on the service – whether IaaS, PaaS, or SaaS – the responsibility for workload and security management may differ between the CSP and the customer.

The cloud is a powerful tool, but as with all new technologies, its relatively recent emergence leaves some challenges, especially in terms of security and trust. As hackers evolve, cloud vulnerabilities are exploited, and misconfigurations remain an area of concern. In the UK, G-Cloud is helping public sector organisations make informed decisions regarding cloud adoption and ensuring high security standards.

Cloud security remains a key issue, and we will undoubtedly see more examples of data leaks and breaches in the future. However, the agility, cost-saving potential, and scalability offered by cloud services make them too valuable for most organisations to ignore.

# Artificial Intelligence (AI) and Quantum Computing

These are two distinct yet revolutionary fields of innovation, each influencing the future of technology in powerful ways. AI aims to mimic human cognition, relieving us of mundane tasks, while quantum computing represents a potential leap in computational speed and architecture.

## Artificial Intelligence

Artificial intelligence (AI) is widely seen as one of the most transformative technologies of our time. The Britannica defines AI as "the ability of a digital computer or computer-controlled robot to perform tasks commonly associated with intelligent beings". The applications of AI are vast and varied, with some key examples including:

- Credit scoring in banking (helping lenders assess the risk of loan applications).
- Automated diagnostics in healthcare.
- Predictive maintenance in manufacturing and industry.
- Judging simple cases in court systems.
- Crime prevention through AI-driven policing.
- Transport optimisation in logistics and supply-chain management.
- Character design in video games, where AI can now create hyper-realistic characters.

In the UK, AI adoption is steadily increasing. The government's National AI Strategy aims to make the UK a global leader in AI by promoting research, fostering public-private partnerships, and ensuring that ethical AI development remains a top priority. The role of AI in the UK is far-reaching – from AI credit risk models in the financial sector to NHS AI Lab, which explores how AI can enhance diagnostic and treatment capabilities in healthcare.

## Augmented Intelligence vs. Artificial Intelligence

At IBM, a concept called "augmented intelligence" was coined to clarify that AI should assist human intelligence rather than replace it. In augmented intelligence, the machine and the human work together, complementing each other's strengths. For example, while AI may perform data-heavy, time-consuming tasks, humans provide critical oversight and nuanced decision-making.

The European Parliament echoes this sentiment by focusing on optimisation, defining AI as "any tool used by a machine to reproduce human-related behaviours, such as reasoning, planning, and creativity". AI tools like Salesforce's Einstein, which analyses customer data to predict customer needs, exemplify this shift toward augmented intelligence. However, AI requires guidance from human operators to ensure that its conclusions are both ethical and accurate.

## AI Bias and Ethical Concerns

The inherent risk of AI lies in algorithmic bias. AI systems learn from the data they are trained on, and if this data is flawed or biased, the AI system will make erroneous decisions. A famous example of this occurred in 2016 when Microsoft's AI chatbot Tay was taken offline after generating racist and inflammatory remarks within 24 hours of its launch. The AI learned from toxic online interactions, highlighting how easily an AI system can go awry without proper safeguards.

In the UK, ethical AI guidelines are being formulated to ensure AI is developed in a manner that aligns with society's values. The Centre for Data Ethics and Innovation (CDEI) has been instrumental in shaping the UK's approach to responsible AI development, providing advice to both the government and industry leaders to mitigate risks, including the use of AI in crime prevention and news generation, which can sometimes produce harmful outcomes such as deepfakes (see Pocket Guide 6).

## AI in Cyber Security

AI has found a prominent place in the cyber security field. Yann Bonnet, Deputy CEO of Campus Cyber in France, notes that cyber criminals are already using AI to make their attacks more efficient and harder to detect. On the other hand, defenders must harness AI to better detect and respond to these sophisticated attacks. In the UK, the National Cyber Security Centre (NCSC) is actively researching how AI can be used to detect, prevent, and mitigate cyber threats.

## Quantum Computing

Quantum computing, while still in the research and development phase, holds tremendous promise for industries requiring ultra-fast and complex computation. Companies like IBM, Google, and Intel are pioneering research in this area, with the potential to revolutionise fields from cryptography to material science.

At the heart of quantum computing is the qubit – a quantum bit that can exist in multiple states simultaneously, unlike classical bits that are either a 0 or a 1. This allows quantum computers to process data exponentially faster than traditional computers.

The UK is actively investing in quantum computing. UK Research and Innovation (UKRI) has outlined significant government funding for the UK National Quantum Technologies Programme, which aims to develop commercial applications for quantum computing across sectors such as healthcare, defence, and telecommunications.

### Cyber Security and Quantum Computing

Quantum computing will have a profound impact on cyber security. Rémi Lassiaille, IBM Technology Managing Director for Vodafone, warned that cyber criminals are already harvesting encrypted data in anticipation of the day when quantum computers can break current cryptographic algorithms. Known as "harvest attacks", these efforts involve stealing encrypted information now, with the hope of decrypting it once quantum computing becomes available.

The UK's National Cyber Security Centre (NCSC) is preparing for a "quantum-safe" future by researching encryption techniques that can withstand quantum computing. The NCSC is advising both private companies and public sector organisations to prepare for the potential threats posed by quantum decryption in the future, with initiatives like Quantum Key Distribution (QKD) helping secure communications.

The innovations driven by AI and quantum computing are pushing the boundaries of what technology can achieve. However, both fields come with inherent risks, particularly around ethical considerations, cyber security vulnerabilities, and the challenges of managing exponentially more complex systems.

## Prospects for 2030

While the cloud, AI, and quantum computing are at the forefront of technological transformation, focusing solely on these innovations can obscure the broader landscape of digital risk. Innovation across industries continues at a staggering pace, and as technology advances, the attack surface for malicious actors widens. Emerging risks associated with this rapid development must be carefully monitored and mitigated.

## Emerging Digital Threats

In 2022, the European Union Agency for Cybersecurity (ENISA) conducted a thorough analysis of the emerging cyber threats that are expected to shape the landscape through 2030. [11] The findings, which resulted in ENISA's Top 10 Emerging Technologies list, highlight how advancements in technology simultaneously present opportunities and threats. Many of these threats arise from the exploitation of new technologies.

Among the most significant threats identified are:

- Supply-chain attacks: With businesses becoming more interdependent and reliant on complex supply chains, supply-chain attacks have emerged as a critical concern. Attackers compromise one vendor, which cascades to multiple companies dependent on that vendor's services or products.

  In the UK, supply-chain attacks are a particular concern for industries like finance and healthcare. In 2022, the UK's National Health Service (NHS) was hit with a cyber attack targeting an external vendor, which disrupted its ability to deliver crucial services across various hospitals and clinics. The National Cyber Security Centre (NCSC) has since issued updated guidance on securing the supply chain, advising companies to establish stronger vendor risk-management protocols.

- Ransomware evolution: Ransomware (see Pocket Guide 1) continues to be a growing threat, with attackers developing more sophisticated techniques to encrypt and exfiltrate sensitive data. The adoption of double extortion tactics – where attackers not only encrypt data but also threaten to release it publicly – has increased the stakes.

  The 2021 ransomware attack on Colonial Pipeline that targeted the US had ripple effects in the UK, highlighting the vulnerability of critical infrastructure to similar attacks. UK energy and transport sectors are taking steps to ensure resilience against such threats, with the Department for Digital, Culture, Media and Sport (DCMS) allocating significant funding to fortify critical systems.

- AI-powered cyber attacks: As artificial intelligence becomes more integrated into cyber security, malicious actors are also leveraging AI to carry out more sophisticated attacks. AI can be used to detect system

---

[11] "Cybersecurity Threats Fast-Forward 2030: Fasten your security-belt before the ride!", 11 November 2022, and "Identifying security threats and challenges for 2030", www.enisa. europa.eu, March 2023.

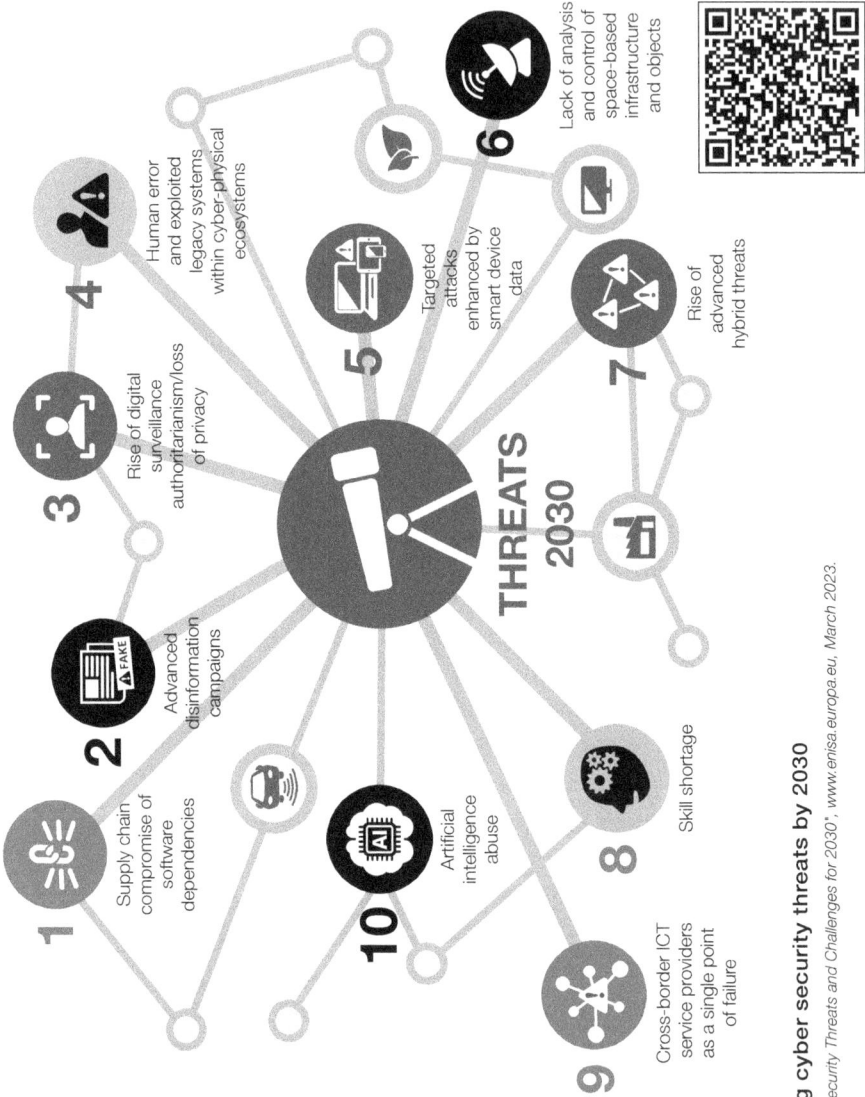

**1** Supply chain compromise of software dependencies

**2** Advanced disinformation campaigns

**3** Rise of digital surveillance authoritarianism/loss of privacy

**4** Human error and exploited legacy systems within cyber-physical ecosystems

**5** Targeted attacks enhanced by smart device data

**6** Lack of analysis and control of space-based infrastructure and objects

**7** Rise of advanced hybrid threats

**8** Skill shortage

**9** Cross-border ICT service providers as a single point of failure

**10** Artificial intelligence abuse

THREATS 2030

EUROPEAN UNION AGENCY FOR CYBERSECURITY

**Figure 1.1** Top 10 emerging cyber security threats by 2030

*Source: ENISA, "Identifying Security Threats and Challenges for 2030", www.enisa.europa.eu, March 2023.*

## Methodology

Through a series of workshops and interviews between March and August 2022, with relevant experts in the PESTLE (political, economic, social, technological, legal, and environmental) fields, ENISA has identified and ranked 21 threats (only the 10 top threats are mentioned here) that will increase in prevalence by 2030.

vulnerabilities, launch phishing campaigns, and automate hacking processes, making attacks faster and harder to detect.

In the UK, GCHQ has warned that AI-powered cyber attacks will become increasingly prevalent by 2030, with the rise of AI systems automating various aspects of cyber threats. The NCSC has begun working on AI-driven cyber security measures to counter these future risks.

■ 5G vulnerabilities: The rollout of 5G networks across the UK presents new opportunities for connectivity, but also introduces new vulnerabilities. The increased number of connected devices, particularly in critical industries like healthcare and transport, means that security measures must be enhanced to protect against potential breaches.

The UK has made significant strides in its 5G infrastructure, with mobile operators expanding networks rapidly. However, as 5G adoption accelerates, Ofcom and the NCSC have focused on ensuring that security standards are incorporated from the outset to prevent potential exploitation by hostile actors.

■ Quantum computing risks: While quantum computing promises to revolutionise industries, it also poses a threat to current cryptographic systems. Once quantum computers become fully operational, they will be capable of breaking many of the encryption protocols that safeguard financial transactions, communications, and national security data.

The UK government has recognised this risk and is investing in quantum-safe encryption methods through its National Quantum Technologies Programme. In particular, the National Cyber Security Centre (NCSC) is advising organisations to begin preparing for the transition to quantum-resistant algorithms, as futureproofing against quantum-based attacks will be critical by 2030.

## Technological Innovations and Risk Management

As technological innovations proliferate, new and unforeseen risks will emerge. ENISA emphasises the importance of a proactive approach to cyber security, particularly as technological boundaries are pushed further. The agency has outlined several steps that governments and organisations should take to mitigate these risks:

■ Adopt Zero-Trust architectures: Organisations are encouraged to adopt Zero-Trust models, where no system or user is automatically trusted. This model focuses on strict verification and minimises the risk of insider threats or external breaches.

In the UK, GCHQ and the NCSC are spearheading efforts to encourage both public and private sector entities to adopt Zero-Trust architectures, particularly in critical national infrastructure sectors such as energy and healthcare.

- Increase cyber security investment: By 2030, cyber attacks are expected to become even more sophisticated, requiring continuous investment in cyber security technologies. The UK has allocated over £1.9 billion to its National Cyber Security Strategy, which focuses on building cyber resilience across all sectors, including energy, finance, and healthcare.

- Enhance cyber security skills development: To meet the growing demand for skilled cyber security professionals, the UK is investing in training initiatives to develop a strong talent pool capable of defending against emerging threats. The CyberFirst programme, led by the NCSC, provides scholarships, apprenticeships, and training opportunities to build the UK's future cyber-defence workforce.

## 2030 and beyond. . .

While digital transformation continues to offer numerous benefits, the associated risks will evolve alongside technological advancements. The role of agencies like ENISA and the NCSC will be vital in ensuring that both public and private sector entities are well prepared to face the emerging threats of tomorrow. By 2030, it is expected that cyber security will be a key driver of both innovation and risk management across all sectors of the UK economy.

# Cyber to Top the Risk List

Cyber risk has become a top concern for organisations across industries, not only because of its technical complexity but also due to the profound business implications it carries. It's often challenging for specialists to convey the full weight of digital risks to non-technical audiences, but recent global events and high-profile breaches have made this issue impossible to ignore. Since 2020, Allianz Risk Barometer, which surveys thousands of risk-management professionals globally, has consistently ranked cyber incidents as the top global business risk. According to the 2022 Allianz Risk Barometer,[12] cyber risk has risen from 40% in 2021 to 44% in 2022, surpassing even traditional business continuity risks. As businesses worldwide rely increasingly on digital systems, the compromise of these systems now poses an existential threat to their operations.

---

[12] "Cyber perils outrank Covid-19 and broken supply chains as top global business risk", Allianz Risk Barometer, www.allianz.com, 18 January 2022.

In the UK, the rise of cyber threats has gained widespread attention, especially in critical sectors like energy, healthcare, and transport. As Raphaël Sanchez, Chief Revenue Officer of Generix Group, aptly describes, "We connect not just physical goods but the associated information flows that drive supply chains. With billions of transactions yearly, any weakness in cyber security can jeopardise the entire system." In the UK for instance, major retailers such as Tesco and Sainsbury's process thousands of transactions daily, making them prime targets for cyber attacks. In 2021, Tesco experienced a significant cyber attack that affected its website and app, disrupting operations and preventing customers from placing orders for over two days. This incident underscores the systemic nature of modern cyber risks.

As organisations digitise more and more of their processes, it's critical that all players – large and small – maintain stringent cyber security protocols. Any single breach could open the door to wider systemic failures, as we've seen with many recent incidents. In Sanchez's words, "It's no longer about just educating employees on the risks; we need to bring everyone in the chain together, ensuring that everyone meets the same standards of cyber defence."

## Homo Digitalis

The primary threat driving the rise of cyber risk is the rapid digitisation of all aspects of modern life. From operational systems to personal lives, nearly every action is now mediated by digital interfaces. During our cyber crisis training sessions, participants often remark on how indispensable IT systems have become to their daily operations. No executive can imagine running a business today without robust IT infrastructure. To highlight the extent of the impact, let's revisit three high-profile cyber incidents that underline the scale and seriousness of modern cyber attacks.

### Maersk – Denmark

On 28 June 2017, Maersk, the Danish shipping giant, was hit by the infamous NotPetya ransomware attack, causing significant operational disruption. Maersk found itself unable to process new orders, and its ability to manage existing operations was severely compromised. The company suffered an estimated $300 million in financial losses. According to Maersk's CEO, the entire infrastructure had to be rebuilt from scratch, which involved reinstalling 4,000 servers, 45,000 computers, and 2,500 applications. This event, which took the company almost two weeks to recover from, highlighted how devastating cyber attacks can be for global

supply chains. Maersk's story became a cautionary tale, underscoring that the scale of cyber risk requires global coordination and exceptional preparedness.

## *Colonial Pipeline – United States*

On 6 May 2021, a ransomware attack on Colonial Pipeline caused the distribution of 45% of fuel along the US East Coast to grind to a halt. Colonial Pipeline operates a vast network of pipelines connecting refineries, airports, and service stations across the eastern United States. The company was forced to halt all its operations, leading to gasoline shortages, price hikes, and widespread panic, with scenes reminiscent of disaster movies as motorists queued for hours at gas stations. President Joe Biden declared a state of emergency on 9 May 2021, highlighting the incident's severity. Colonial Pipeline eventually paid a $4.4 million ransom in Bitcoin, though 80% of this was later recovered by the US Department of Justice. This incident demonstrated how vulnerable critical infrastructure can be to cyber attacks, raising concerns not just in the US but globally.

## *NHS – United Kingdom*

In May 2017, the NHS was severely impacted by the WannaCry ransomware attack, which affected more than 80 healthcare institutions. The attack encrypted data and demanded a Bitcoin ransom, causing widespread disruption. Thousands of appointments and surgeries were cancelled, emergency departments shut down, and ambulances were diverted. Although no direct fatalities were reported, the delays in care raised significant concerns. The National Audit Office later criticised the NHS for inadequate cyber security measures. In response, the UK government invested £210 million in improving NHS cyber security.

In the UK, this incident prompted the National Cyber Security Centre (NCSC) to re-evaluate the cyber security posture of the UK's critical infrastructure, including energy, water, and transport sectors. The UK government introduced stricter cyber security measures for critical systems, recognising that attacks like the one on Colonial Pipeline could happen on British soil, with devastating consequences.

# A Growing Surface Area

Cyber attacks have become so common that scarcely a week goes by without a headline about a major breach affecting either private companies or public institutions. The frequency and scale of these incidents have become so ingrained in popular culture that numerous TV shows and films revolve around cyber crime as a primary plotline – *NCIS Los Angeles*, *CSI Cyber*, *Mr. Robot* – and others explore cyber crime as an existential threat to society. Yet no company executive wants to be featured in the media as the latest victim of a cyber attack, knowing it could result in damaged reputation, operational shutdowns, or even legal consequences.

This ubiquity of cyber attacks stems from our growing dependency on technology. The very act of digitising operations – whether through cloud computing, the Internet of Things (IoT), or artificial intelligence – has expanded our digital vulnerabilities. By putting more data online and relying more heavily on interconnected systems, businesses are unwittingly increasing their risk of being targeted. Thomas Billaut, Head of Cyber Operations at Forvia, aptly describes this phenomenon: "Cyber is a frantic race to win business, it's already going at high speed and it's accelerating every day."

## Imperfection Is Intrinsic to IT

While cyber risks have always existed, they are exacerbated by the inherent flaws in IT and software systems. Human error is frequently cited as a leading cause of cyber incidents, but the second major vector for attacks is the vast array of imperfect software programs we use daily. Every patch, update, and new feature release serves as a reminder that the code is intrinsically flawed. Microsoft's decision to make automatic updates mandatory in its latest Windows release reflects this reality – security updates have become so critical that user intervention is no longer optional.

This scenario is akin to buying a car without brakes, only to be told they'll be installed in a few weeks. It sounds absurd in any other industry, but in software, it's standard practice. Not all software is created equal in terms of security. Programs developed for sensitive applications, such as military or cyber security systems, often have more stringent security protocols. However, the level of security is frequently dictated by consumer demand. For instance, how many consumers check the digital security features of a coffee machine or a webcam before purchasing one? In 2020, hackers exploited a vulnerability in a French connected coffee maker to gain access to sensitive systems, and in 2018, cyber criminals attacked American casino security cameras. Such incidents underscore the point that anything connected to the internet is a potential target.

## Software Code Is Inherently Imperfect

The imperfection of software code is a recognised reality in the industry. As companies collect and process more data, the potential for a cyber attack grows. IBM's 2020 Global C-Suite Study[13] highlighted a significant shift since COVID-19, showing that digital transformation has accelerated rapidly across industries, bringing both innovation and inherent risk. With data often described as the "new oil" – a highly valuable asset – businesses are increasingly hoarding data for insights, personalisation and efficiency. However, this wealth of data also makes them prime targets for cyber criminals, whose attacks on this critical asset have become more frequent and more sophisticated.

In the UK, this is particularly evident with regulations such as GDPR (General Data Protection Regulation), which imposes strict penalties on organisations that fail to protect user data. A breach not only results in financial losses but also severe reputational damage. UK organisations, both public and private, are increasingly aware of these vulnerabilities as they digitise their operations and rely more heavily on data-driven decision-making.

## The Expanding Cyber Exposure

The increasing integration of IT into all facets of business and personal life has expanded the "attack surface" – the total number of points where an attacker can attempt to gain access to an IT system. Pierre Chaffardon, Managing Director France of Generix Group, articulates the dilemma well: "IT represents two sides of the same coin when it comes to security: unprecedented computing power and analytical capabilities can enhance security, but those same tools can be used by attackers to penetrate increasingly interconnected systems."

> *"It's no longer a question of 'if' an attack will happen, but 'when'."*

The democratisation of cyber-attack frameworks, where hacking tools are marketed and sold as readily as Software as a Service (SaaS), compounds the threat. On the dark web, everything from ransomware toolkits to zero-day exploits is available for purchase, turning cyber crime into a service industry of its own. Consequently, the responsibility falls not only on individuals and companies but on management committees to stay informed, continually evaluate risk, and prepare for inevitable attacks. As Chaffardon observes, "It's no longer a question of 'if' an attack will happen, but 'when'."

---

[13] "Majority of global C-suite executives are rapidly accelerating digital transformation due to COVID-19 pandemic, but people and talent are key to future progress", IBM Global C-Suite Study, IBM Newsroom, 30 September 2020.

# Cyber Security: Anecdotes and Lessons

One humorous yet telling anecdote from a 2021 CIO conference illustrates the seriousness with which cyber hygiene should be taken. During a Q&A session, an attendee asked what they should do if their chairman found passwords inconvenient, wanted open Wi-Fi access, and insisted that USB ports remain active on all company computers. Our response? "Look for a new job immediately." While the scenario was meant to evoke a chuckle, it underscores the reality that the level of cyber security in an organisation is ultimately a management decision, and a poor one can lead to disastrous consequences.

The analogy of protecting digital assets is like guarding a jar of cookies. You wouldn't leave the jar on the counter for anyone to access freely. Instead, you'd place it out of reach or lock it away. The same principle applies to cyber security – firewalls, encryption, and monitoring are necessary to ensure that no one can simply "grab" your data without consequences.

Cyber security in the UK has become a cornerstone of risk management across all industries, from finance to healthcare. As the UK government invests more in national cyber security initiatives, it remains critical for private companies to follow suit and guard their growing digital surfaces with the same vigilance. The cyber race isn't slowing down – if anything, it's accelerating as companies compete for digital supremacy in an increasingly risky environment.

---

**Key Takeaways**

To say that our world is going digital is an understatement. Just as we can't fight against the changing seasons, we must adjust our strategies, tools, and mindset to make the most of the digital reality we're now a part of. Embracing digital transformation is no longer optional, it is essential. Like in a marriage, we have to accept the digital world "for better or for worse," with all its advantages and inherent risks.

As we prepare to dive into the next chapter, which examines the business model of cyber crime, let's summarise the key takeaways from this chapter:

■ **Daily digital interactions**
  Digital and connected devices have become an integral part of our everyday lives. From coffee machines and televisions to the cars we drive, digital technology is all around.

▶

- **Limitless innovation**
  The only boundaries to digital innovation are human imagination and technological capacity. While we once relied on 56 KB modems, today we are pushing the limits with quantum technology. The digital age is driving improvements in quality of life, environmental sustainability, and health care, offering us tools that were unimaginable a few decades ago.

- **Increased cyber risks**
  As our dependence on digital technology grows, so too does the surface area for cyber attacks. Safeguarding our digital well-being isn't just a matter of convenience, we have a duty to protect ourselves from the rising threats that this technological evolution brings.

If we recognise these dynamics, we can strike the right balance between seizing the opportunities of the digital age and mitigating its risks.

# 2

# THE CYBER CRIME BUSINESS MODEL

"To effectively protect yourself from criminals,
you have to think like a criminal."

– Lee Morrison

In this chapter, we'll explore the inner workings of cyber crime, examining the numbers, organisation, business models, guerrilla warfare-like tactics, and the collaboration between threat actors. More than that, we'll focus on a mindset that mirrors the world of criminals themselves – a necessary skill for anyone who is serious about protecting themselves against the growing cyber threat.

Imagine the world's economic losses from cyber crime being equivalent to six subprime crises every year. The scale is staggering, but that's the reality we are facing.

A while ago, we posted a quick note on LinkedIn that said, "If Pablo Escobar were still with us, he would have swapped his cocaine empire for a computer." While it may seem like we're being provocative and poking the prover-bial "Cocaine-Bear", the comparison is accurate: cyber crime is not a niche phenomenon, it's the new underworld economy. Internet service provider Beaming reported that in 2023, UK businesses lost £31.5 billion to cyber crime, more than double the figure from 2019. Imagine being hit with these kinds of losses every year. That's the cost.

Globally, Cybersecurity Ventures predicted in 2020 that cyber crime will cost the world $10.5 trillion by 2025. If it were a country, it would be the third-largest economy in the world, trailing only behind the US and China. These numbers aren't just statistics, they point to a new business model that thrives in the shadows of the digital world. Again, I am no mathematician, but those figures seem rather high.

How does this parallel world operate? Who are the players? Why can't we seem to stop it? To answer these questions, we need to adopt a mindset not unlike our attackers.

In the world of physical security, for example, safe designers have been known to employ former safe crackers – people with intimate knowledge of how to break into them – to build more secure systems. This principle has long been effective because nobody understands the weaknesses in a system better than those who have exploited them. The same is true in digital security. In fact, the FBI famously employed Frank Abagnale Jr., a con man whose life was later depicted in the film *Catch Me If You Can*, to help them spot fraud and counteract confidence scams.

Just as physical security experts use former criminals to build stronger defences, the fight against cyber crime requires a shift in how we think about protection. It demands not just the best technology but a mindset that can anticipate, outwit, and neutralise threats. In this chapter, we'll look at how cyber criminals operate like a business – albeit one built on deception, exploitation, and even collaboration. But just as criminals innovate, so too must we.

As we uncover the business model behind cyber crime, remember it's not about stopping every attack, but understanding how these modern-day criminals work so we can stay one step ahead.

# Dial M for Malware: An Unavoidable Threat

Every year, IBM Security publishes its "Threat Intelligence Index" on the state of cyber threats from the past year. The latest figures are very informative, particularly regarding the sectors most frequently under attack. Figure 2.1, adapted from the report, shows the breakdown of attacks by business sector and illustrates that, for the first time since records began, the finance industry has been knocked off the top spot.

To understand why all sectors come under attack – but not with the same intensity – we need to follow the money. Over 90% of cyber attacks are motivated by financial gain; the rest are either state-sponsored attacks designed to destroy or collect intelligence, or ego-fuelled attacks to prove a point. Cyber-criminal groups are looking for a return on investment; they want the outcome to be worth their while.

If you were a cyber criminal, what would you do? Would you attack an impregnable fortress with little chance of success? Nor would we. The same applies to cyber criminals. If your digital infrastructure is critical to the

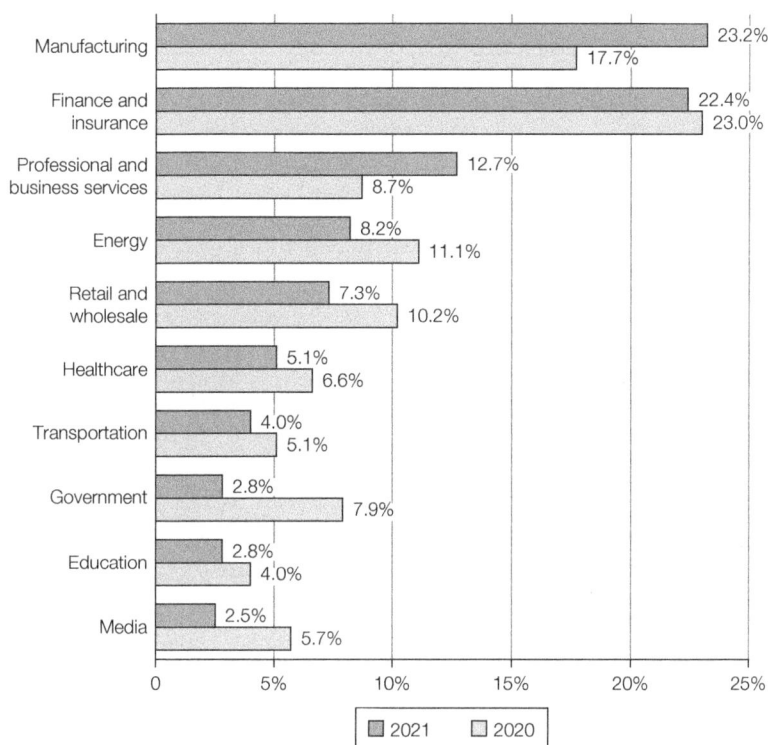

**Figure 2.1** Breakdown of attacks by sector between 2020 and 2021

*Source: IBM Security X-Force, 2022.*

running of your business, if your defences aren't up to scratch, and if you've got information that's worth reselling on the black market, you're in the line of fire. That's when targeted attacks happen. For non-targeted attacks, cyber criminals exploit vulnerabilities in widely used systems (that's the beauty of globalisation, the cloud, and the cross-industry factorisation of systems and software). This technique can best be described as a message in a bottle. Let's not forget that the companies hit by NotPetya in 2017 (Maersk, Renault, Village Doctors, etc.) were convenient collateral damage from a cyber attack that the Russians created to target Ukraine. . . And that's still all too relevant today.

We often use another example to illustrate the thought process of a cyber criminal (i.e. what we would do if we were in their shoes). Why were hospitals in so many countries targeted in 2021 and 2022? Well, what happens in a hospital? Is it a world where peace reigns, budgets are generous, and cyber security is a priority? Of course it isn't. But you might wonder why hospitals are attacked if they don't have the money to pay a ransom when their information systems are blocked, or patient data is stolen.

In the UK, for example, hospitals have found themselves increasingly in the crosshairs of cyber criminals. In 2021, the National Health Service (NHS) in Northern Ireland was severely disrupted by a ransomware attack that targeted the Health Service Executive (HSE). The attack forced healthcare facilities to revert to manual processes, causing delays in patient care, appointment cancellations, and a temporary shutdown of IT systems. Like many healthcare providers worldwide, the NHS has found itself struggling to balance tight budgets with the need to invest in adequate cyber security measures.

As referenced in Chapter 1, this attack highlighted the vulnerability of the healthcare system to cyber threats, and while no ransom was paid, the financial and operational toll was significant. It underscored the need for better cyber security measures in healthcare, just as the WannaCry attack had done in 2017. In response to these threats, the UK government allocated substantial funds to bolster cyber security in the NHS as part of its long-term digital health strategy, including a £210-million investment to strengthen IT security and resilience across the healthcare system.

When hospitals are under-resourced and management has other priorities, cyber security often falls by the wayside. In the case of the NHS, like other healthcare systems globally, a mix of ageing infrastructure, limited resources, and constant operational pressure makes it an attractive target for cyber criminals. The result? Major disruptions to critical services, which, in extreme cases, can even lead to loss of life.

For instance, in Germany in mid-2020, a person died in Düsseldorf University Hospital after a cyber attack crippled the hospital's systems, delaying vital treatment. While death is part of hospital life, this was the first known case of a fatality directly linked to a cyber attack.

It's not the number of attacks that counts; it's the impact of the ones that get past the lines of defence.

The other point to highlight in Figure 2.1 is the systematic inclusion of finance among the two most-attacked sectors. Paradoxically, this ranking does not include the most affected companies. Yes, there are exceptions, such as the $40 million ransom paid by CNA Financial in the United States in May 2021. But that remains an isolated case; other industries have been more severely affected.

Again, it doesn't matter how many attacks are made. What's important is the impact of an attack. So why is the finance sector so heavily targeted, but not as badly affected? The answer is simple: it was attacked first ("let's rob a bank!"), so it started protecting itself first. Remember that IT is the production tool for a bank or an insurance company, so by its very nature, this sector pays particular attention to its defences. This leaves the pharmaceutical, energy, or transport sectors more vulnerable.

Another interesting phenomenon emerged from the 2021 edition of IBM's annual report. Why, in 2019, were manufacturing and energy only the eighth and ninth most attacked industries respectively, but by 2020, they had moved up to second and third place? The answer can be summed up in five letters and two numbers: COVID-19. Did people travel much during that time? No. Did we do a lot of online shopping from home? Yes. Is it more profitable to attack companies in a sector that has money during a time like that, or companies in a sector that is struggling? These are rhetorical questions, but you see the point.

The same applies to seasonality. If you're interested in the toy industry, does it make more sense to launch an attack around Christmas, when the company probably makes 80% of its annual sales, or in the middle of May? In any case, whatever the sector, all companies depend on IT. If you are in a competitive market, you are a target.

To finish this section on targets, look at Figure 2.2 from the 2022 IBM Security report to see that it's not just about business sectors. Yes, all sectors are affected, but so are all geographical areas. It's also interesting to note that in 2019, North America was the most attacked, then in 2020 it was Europe, and in 2021, Asia. It's a game of cat and mouse, and nobody is safe.

## Who's Really Behind the Hoodie?

When you hear the word "hacker", the mind often races to an image of a hooded figure hunched over a laptop in a dim basement. Google it, and that's the very image you'll see: a shadowy, lone figure working away with malicious intent. But that stereotype misses the mark. Hackers today are as diverse in their motivations as they are in their skills.

In reality, hacking is just a skill set, a tool. There are those who wield it for good, and they are known as "white hat" hackers, working to strengthen systems and protect data. On the flip side, "black hats" leverage the same skill to exploit and steal, often for personal or financial gain. To avoid confusion, we'll refer to these malicious actors as "threat actors" or "cyber criminals".

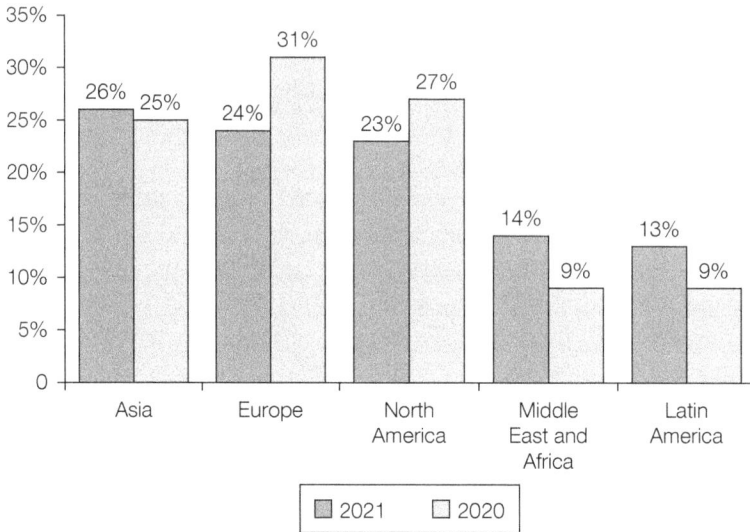

**Figure 2.2**  Breakdown of attacks by geographical area between 2020 and 2021

*Source: IBM Security X-Force, 2022.*

Hackers today are as likely to work for major corporations or governments as they are to operate independently. The old basement-dweller image has given way to highly organised teams, some state-sponsored, working together in a web of digital espionage and theft.

Interestingly, the term "hacker" has been reclaimed and even embraced by the cyber-security industry as a positive force. Companies like IBM have leaned into the word to market their cyber security capabilities, using phrases like "hack-driven offense" to describe their proactive approach to cyber defence. IBM's "Offensive Security" division, for instance, uses ethical hackers to simulate attacks and stress-test a company's systems. This shift in language reflects the growing recognition that to truly protect digital systems, you need to think like a hacker, finding vulnerabilities before the bad guys do. It's a spin on a word that once had entirely negative connotations, showing how even the world of cyber security is evolving to embrace the hacker mindset as a tool for good.

This dual nature of hacking highlights a key point: hacking itself is not inherently good or bad, it's simply a skill set. How that skill set is applied determines its ethical standing. Just as ethical hackers work to strengthen security, cyber criminals exploit those same vulnerabilities for malicious purposes.

Understanding both sides of this coin is crucial for developing strong defences against cyber threats.

## What Drives Threat Actors? Lambo's!

Understanding what motivates these threat actors is key to defending against them. To beat an opponent, you need to know what they want. While financial gain is the primary motivator behind more than 90% of cyber attacks, there are other reasons hackers engage in such activities. Here's a breakdown of the most common motivators:

■ Money: As mentioned earlier, financial gain is the leading driver behind most cyber attacks. Ransomware, intellectual property theft, and sensitive data breaches are all means by which criminals extract value from their victims, usually transacting in cryptocurrencies to remain anonymous. In many cases, organised crime networks use cyber crime to fund their other illicit activities. For example, the UK's National Crime Agency has observed an increasing overlap between cyber-criminal groups and traditional organised crime, with the Dark Web playing a key role in laundering proceeds.

■ Political Motivation: State-sponsored cyber crime has increasingly become a tool for geopolitical influence. Governments may employ hackers to disable critical infrastructure, manipulate elections, or carry out ransomware attacks to create confusion and chaos. The 2017 WannaCry ransomware attack, which severely impacted the UK's National Health Service (NHS), was allegedly tied to North Korean actors, illustrating how state-backed cyber operations can have far-reaching consequences. This attack disrupted hospital systems, delayed surgeries, and compromised patient care, showing how political motives can devastate vital public services.

■ Corporate Espionage: Compromising a competitor's information system offers a treasure trove of potential benefits. These range from stealing trade secrets or intellectual property to blackmailing the company with insider knowledge. Gaining access to a competitor's sensitive data could damage their reputation, force them to divert resources away from core business activities, and even lead to significant legal trouble, especially if the breach results in the exposure of personal data. While this kind of attack is typically state-sponsored or backed by nation-states, some rogue businesses may also engage in it, particularly in regions with lax enforcement of international law.

▪ Ideological or Activist Causes: Some hackers operate for ideological reasons, believing their work serves a higher purpose. Hacktivist groups like Anonymous target corporations and governments they believe are corrupt, often with the aim of exposing immoral or illegal activities. In the UK, notable hacktivist actions have included DDoS attacks against financial institutions to protest economic inequality and actions against public institutions for perceived social injustice.

▪ Revenge or Spite: Not all cyber criminals are in it for the money. Some attacks are driven by personal grievances or revenge. A dissatisfied employee, for example, might launch an insider attack after being laid off or denied a promotion. Similarly, a customer upset with a faulty product might decide to retaliate by exploiting vulnerabilities in the company's digital infrastructure. The infamous TalkTalk breach in 2015 involved a teenager who accessed sensitive customer data simply to prove that he could. This was a non-financially motivated attack that still caused immense damage to the company's reputation and finances.

# Digitisation and Its Discontents

To fully understand the threat landscape, it's critical to examine the groups behind the attacks. As mentioned earlier, CNA Financial paid a ransom of $40 million in 2021, but they weren't the only ones. In the same year, global meat-processing giant JBS paid an $11 million ransom, while Acer faced a demand for a staggering $50 million. At this point, it's fair to ask, "How could a teenager operating from a garage launder such massive sums?" The answer is simple: they're not the ones behind these attacks. Today, most cyber attacks are carried out by highly sophisticated organisations that function much like legitimate businesses, albeit with far more dubious goals. And in many cases, law enforcement agencies, particularly in parts of Eastern Europe, Asia, and Latin America, aren't exactly cracking down on these groups with the same vigilance found elsewhere.

## "FriYay!"

Consider the observation from our colleague at Symantec, who noted that their system consistently blocked more cyber-attack attempts on Friday evenings than at any other time during the week. Why? Simple. Cyber criminals run operations like any other business. Employees clock in on Monday morning and clock out on Friday evening. Therefore, to maximise weekend downtime, many cyber gangs cast wide nets on a Friday night, hoping for a good "catch" over the weekend. By Monday morning, they can reel in the compromised systems and start extracting the data or deploying ransomware.

A telling example of this structured operation came from the international summit against ransomware, held by the White House in October 2022. Thirteen technology companies and around 30 nations attended, including the UK, France, Italy, and Germany. Notably absent from the discussions: Russia. It's no coincidence, given that 75% of the 793 ransomware incidents reported to US authorities in the second half of 2021 were linked to Russian agents or those operating on Russia's behalf. This illustrates that cyber-criminal syndicates often operate with the tacit approval, or even protection, of their home countries – as long as they avoid targeting domestic interests.

Take Maksim Yakubets, leader of the notorious Russian cyber crime syndicate Evil Corp. Far from being a reclusive teenager hacking away in his parents' basement, Yakubets reportedly flaunts his wealth, owning luxury cars like Lamborghinis. Evil Corp operates as a well-oiled machine, resembling a corporation more than a gang of opportunistic hackers. This example highlights how cyber crime has evolved far beyond lone wolves or isolated groups; it is now a global, organised industry.

## Teamwork Makes the Nightmare Work

If cyber security is a team sport, so is cyber crime. Cyber-criminal groups collaborate and compete, much like corporations. Their operations are well-coordinated and, in some cases, highly professional. This is precisely why cyber attacks have become so difficult to counter and why preparation is crucial.

Today, cyber-criminal organisations even have customer service call centres for ransomware victims. Yes, you read that correctly – a human operator at the other end of the line to help the victim navigate the ransom payment process. Welcome to the bizarre world of cyber crime.

Behind the eccentric names – Clop, Conti, REvil, Ryuk, Sodinokibi – are highly organised structures. Let's break down the five main roles identified within cyber crime syndicates:

- ■ Advanced Persistent Threat (APT) Groups: These are state-sponsored actors or groups that mix espionage with financial crime. They often operate under the protection of their government, targeting industries for sabotage, intelligence gathering, or financial gain. APT groups often develop custom malware with sophisticated surveillance capabilities. A notorious example of an APT attack is the Stuxnet worm, which was designed to disrupt Iran's nuclear programme by damaging centrifuges at a facility in Natanz.

- Malware Developers: These actors create malicious software that exploits known and unknown vulnerabilities (zero-days). Historically, they were highly skilled hackers, but as the cyber crime industry has matured, malware development has become more professionalised, with some groups operating like software development firms – complete with regular updates and customer support!

- Information Brokers: These criminals trade in personal data (names, addresses, social security numbers, account details) often selling the information on dark web marketplaces. The scale of these operations can be staggering. In the Equifax breach of 2017, the personal data of 143 million Americans was stolen and later sold on the dark web. Similarly, the Yahoo! breach in 2013 compromised three billion user accounts.

- Initial Access Brokers (IAB): Specialising in selling access to compromised systems, IABs play a crucial role in enabling large-scale attacks. These actors sell login credentials and other access points, making it easier for cyber criminals to launch deeper, more devastating attacks, like corporate espionage or ransomware.

- Ransomware as a Service (RaaS) Providers: A recent development in the world of cyber crime, RaaS operates like any other Software-as-a-Service (SaaS) model. These groups sell ransomware toolkits and infrastructure to cyber criminals looking to carry out attacks. RaaS is less risky than collecting ransom payments directly, as the providers simply supply the tools and take a cut of the proceeds.

# Organised Cyber Crime

As you can see, the world of cyber crime is highly structured, with specialisation across different roles allowing for increasingly sophisticated attacks. The collaboration between threat actors is striking. In a recent case observed by Orange Cyberdefense, a major French cyber security company, groups from Brazil and Russia were seen working together. The Brazilians used brute-force techniques to crack passwords, while the Russians employed social engineering methods to impersonate senior executives in "president fraud" scams (see Pocket Guide 1).

This level of cooperation is nothing new in the criminal underworld. Cyber criminals are just following the age-old principle of specialisation: some produce, others distribute. In the digital age, this coordination allows cyber-crime syndicates to operate on an industrial scale, making them a formidable opponent.

Cyber criminals have evolved from lone operators to highly organised, sophisticated enterprises. Just as businesses in the legitimate economy cooperate to maximise profits, so too do cyber-crime groups, pooling their skills and resources to devastating effect. The fight against them requires an equally coordinated and adaptive response.

## The World of Cyber Guerrilla Warfare

The cyber threat landscape is vastly fragmented, making it notoriously difficult to contain. Cyber criminals collaborate, but they also compete, constantly jockeying for control over different sectors of cyber space. When companies find themselves under attack, they often mistakenly believe they're dealing with a single threat actor. Multiple groups – sometimes coordinated, sometimes not – may be launching simultaneous or cascading attacks. Some companies, tragically, are attacked repeatedly, even after paying a ransom. This revolving door of exploitation stems from a simple truth in cyber crime: once a payer, always a payer.

A threat from a single, identifiable source is challenging enough. But when cyber attacks originate from autonomous hackers operating from any corner of the globe, shielded by their home country's authorities or working as state-sponsored actors, the situation becomes far more complex. Some cyber criminals enjoy the tacit protection of their governments, so long as their targets are foreign entities or adversaries. In such a context, adopting a clear and effective defence strategy can feel like battling an unseen, shapeshifting enemy.

## Don't Show Them the Money. . .

Let's get back to "once a payer, always a payer" in the world of ransomware. Paying a ransom might restore access to a company's data temporarily, but it also paints a target on its back, signalling to other cyber criminals that this organisation can be extorted again. Cyber criminals thrive on this cycle of ransom payments and reinfection, perpetuating a vicious loop that's hard to break.

In its 2021 "Global Threat Report", CrowdStrike presented a compelling diagram (Figure 2.3) illustrating the interwoven but highly atomised nature of cyber threats. It highlights how every group plays a distinct role in the larger web of cyber warfare, showing that those pulling the strings may not be the ones launching the attacks.

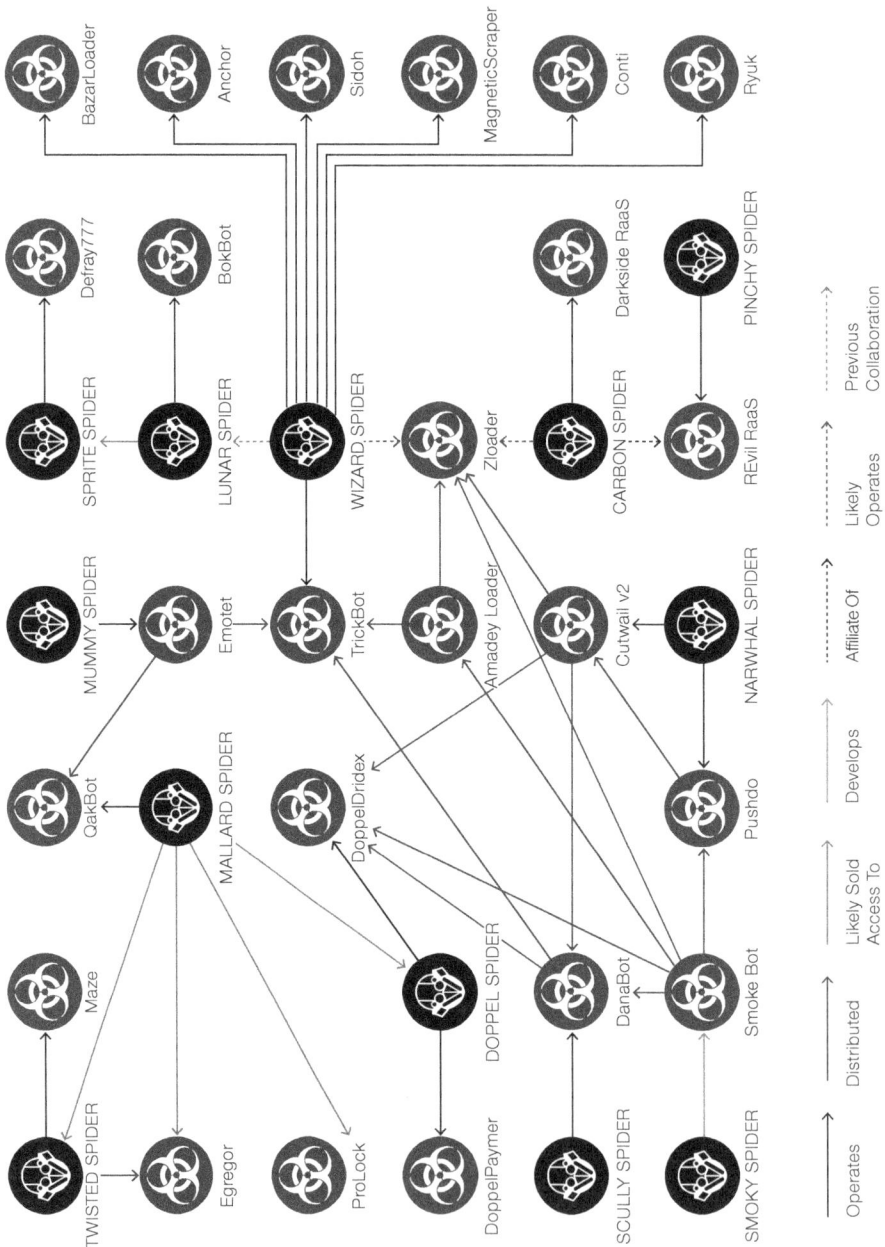

**Figure 2.3**  Group of attackers and interactions

*Source: CrowdStrike, 2021 "Global Threat Report", www.crowdstrike.com.*

Legend (bottom):

Previous Collaboration · Likely Operates · Affiliate Of · Develops · Likely Sold Access To · Distributed · Operates

Much like traditional guerrilla warfare, the cyber underworld consists of both minority and majority groups. These groups can be classified by their effectiveness and their global impact. It's generally the most successful majority groups with familiar names who cause significant damage to organisations and systems across the world. Some of the most notorious ransomware groups include REvil, Conti, and DarkSide, whose attacks have made global headlines by crippling vital industries, from fuel pipelines to food processing giants.

## The Major Ransomware Players

*"Infamy, infamy they've all got it in for... you!"*

The world of ransomware has seen several key players rise to infamy, orchestrating high-profile attacks across the globe. These groups employ sophisticated techniques and leverage the Ransomware as a Service (RaaS) model to broaden their reach, allowing affiliates to carry out attacks in exchange for a cut of the ransom profits. Each of these groups has a distinct modus operandi, but the common denominator is their effectiveness in causing widespread disruption and demanding exorbitant ransoms.

### Avaddon

The Avaddon ransomware first appeared on the radar in 2019, quickly becoming a significant threat. Like many other ransomware variants, it is sold as a Ransomware as a Service (RaaS), allowing cyber criminals with less technical expertise to launch attacks by using Avaddon's infrastructure. A notable victim of an attack involving Avaddon ransomware was AXA, the French insurance giant, in May 2021. During the attack, threat actors stole customer credentials, contracts, reports, and other sensitive information. The attack came just days after AXA announced it would stop covering ransom payments in its cyber insurance policies, possibly triggering the attack as retaliation.

Avaddon is typically spread through phishing and spam campaigns that contain malicious attachments or links. Once the ransomware is deployed, it locks users out of their systems, demanding a ransom in cryptocurrency. However, after a spree of attacks in 2021, the group behind Avaddon suddenly announced their withdrawal from the scene, releasing decryption keys for their victims, though the extent of their operations remains a subject of speculation.

## Clop

Originating in Ukraine, Clop ransomware emerged in early 2019 and quickly gained notoriety for its brazen attacks and massive ransom demands. One of the most significant examples was its assault on Software AG, a German software company, in 2020. The attack included a ransom demand of $20 million and exposed the company to severe data breaches. Other victims included the US bank Flagstar, security-software company Qualys, and academic institutions like the University of Colorado and Stanford University.

What sets Clop apart is its ability to adapt to different environments. In several cases, it targeted file transfer solutions like Accellion's File Transfer Appliance (FTA), allowing it to steal large troves of data from organisations before encrypting their files. This tactic has created additional leverage for the group, as they can threaten to release sensitive data publicly if the ransom is not paid.

## Darkside

One of the newer but highly impactful players on the ransomware scene is Darkside, which emerged in the third quarter of 2020. Darkside operates under the RaaS model, enabling affiliates to use their ransomware to attack targets, often splitting profits with the group. Their most infamous attack to date was the assault on Colonial Pipeline in May 2021. The attack led to the shutdown of the largest pipeline in the United States, which supplies nearly 45% of the fuel used on the East Coast. This caused fuel shortages, price spikes, and widespread panic buying, underscoring the vulnerabilities in critical infrastructure.

Although Darkside was relatively new to the ransomware landscape at the time of the attack, their professionalism was evident. After causing such large-scale disruption, Darkside posted an apology on their website, stating that their intention was to make money, not to create chaos in society. Despite the apology, the Colonial Pipeline attack had lasting repercussions, highlighting how ransomware can have far-reaching effects on the economy and everyday life.

## REvil/Sodinokibi

The REvil (also known as Sodinokibi) ransomware group rose to prominence in 2019, quickly becoming one of the most notorious ransomware syndicates. REvil's operations follow the RaaS model, enabling less experienced hackers to carry out attacks using their ransomware. The group gained global attention in 2021 when it targeted Acer and Quanta Computer, demanding $50 million in ransom from each. Quanta Computer, a key Apple supplier, had its sensitive engineering plans and data stolen, putting immense pressure on the company.

REvil was also responsible for orchestrating the attack on Kaseya in July 2021, which led to the encryption of thousands of systems across hundreds of companies, making it one of the largest ransomware attacks in history. The Kaseya attack was so significant that it prompted action from the US government and led to a temporary shutdown of REvil's dark web presence. However, REvil re-emerged, showing just how resilient and persistent these groups can be.

## Ryuk/Conti

First appearing in 2018, Ryuk ransomware made headlines for targeting high-value institutions in the US, UK, and France. Among its most notable victims were Jackson County, which paid a $400,000 ransom, Riviera Beach, which was forced to pay $594,000, and LaPorte County, which paid $130,000. Ryuk's operators have proven particularly adept at targeting local governments, healthcare providers, and educational institutions, exploiting the fact that these entities often cannot afford extended downtime.

Conti, another variant closely linked to Ryuk, continued this legacy, amassing a substantial number of victims. One of its high-profile attacks targeted the Health Service Executive (HSE) in Ireland in May 2021. The attack caused severe disruption to Ireland's healthcare system, affecting diagnostic services, patient appointments, and more. The attack on HSE demonstrates the destructive potential of ransomware when critical public services come under fire, underscoring how ransomware groups like Ryuk/Conti prioritise chaos to extract larger payouts.

According to the IBM Security X-Force report, the last two ransomware groups alone (Figure 2.4) accounted for half of all ransomware attacks worldwide in 2021. Trends, it seems, are not just the prerogative of the music industry, and unlike a catchy tune, we can't just *shake them off*.

One of the key challenges in cyber defence is that it often feels like an asymmetrical battle – where the playing field is far from even. While larger organisations may employ hundreds or even thousands of cyber defenders, from in-house teams to contractors, they are still being bested by just a few cyber criminals. Despite having significant resources, David is consistently crushing Goliath in the world of cyber warfare. But why does this happen so often? Let's examine the reasons:

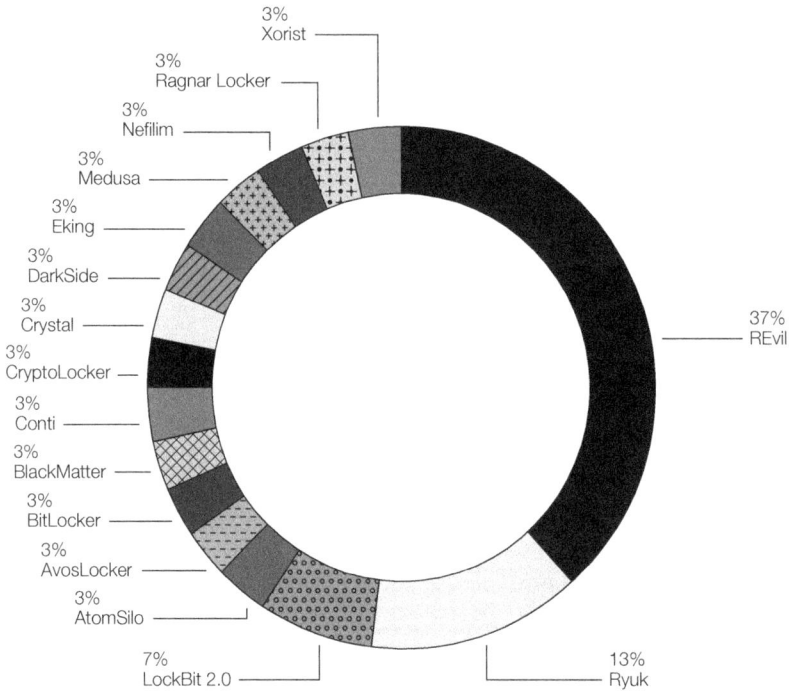

**Figure 2.4**  **Types of ransomware observed in 2021**

*Source: IBM Security X-Force, 2022.*

Pie chart labels:
- 3% Xorist
- 3% Ragnar Locker
- 3% Nefilim
- 3% Medusa
- 3% Eking
- 3% DarkSide
- 3% Crystal
- 3% CryptoLocker
- 3% Conti
- 3% BlackMatter
- 3% BitLocker
- 3% AvosLocker
- 3% AtomSilo
- 7% LockBit 2.0
- 13% Ryuk
- 37% REvil

## Outnumbered and Outgunned

In sports, it's a widely accepted principle that the team or player who strikes first often has the upper hand. However, cyber defenders almost always find themselves in a reactive position. They stay informed about emerging threats (as we'll discuss in Chapter 3), but the sheer volume of potential risks makes it nearly impossible to be fully offensive and proactive. To put this into perspective, during a recent training session with a senior leadership team at a US-based insurance company, the CISO told us their security team handles over 3 million attacks per month. Yes, per month.

## Keeping a Clean Sheet

The asymmetry here is stark. Attackers only need to be successful once, while defenders must be perfect every time. Think of it like football: if a goalkeeper makes one mistake, the opponent scores, and the game could be lost. But if the striker misses a shot, there's always another chance later in the match. In cyber defence, a single mistake can mean a catastrophic breach, yet defenders must stop every attack, no matter how small or seemingly insignificant.

## Hamstrung Counterattacks

In many countries, including the UK, launching a counterattack is illegal, even if it might seem justified. Legally, the principles of self-defence might apply, but the situation becomes murky when attackers hide behind compromised third-party devices. Imagine this: your grandmother's laptop, your neighbour's smart fridge, or your friend's connected car is used as a bot in a cyber attack. Are you going to launch a counterstrike to shut down your neighbour's fridge? Clearly not.

## Stalemate Not Checkmate

Even in countries where offensive cyber actions are permitted, they are rarely used. cyber-defence expertise is in short supply globally, and budgets for cyber security are far from unlimited. Even basic measures can be hard to maintain consistently, leaving little room to mount an offensive. Additionally, engaging in cyber "combat" against a well-trained, professional attacker is risky. In many cases, it's more pragmatic to focus on fortifying defences and diverting attackers, much like the old survival strategy: you don't have to outrun the lion; you just have to outrun the other guy.

## The Hydra Problem

Battling cyber crime is like facing the Lernaean Hydra; cut off one head, and several more emerge in its place. This is particularly true given the enormous potential rewards for cyber criminals. Even when law enforcement agencies make significant busts, such as Europe's recent efforts with Interpol including Operation Haechi III during which $130 million was seized from online fraudsters, the victory is often short-lived. New threat actors quickly take the place of those who have been captured or neutralised. The enemy is agile, intelligent, and constantly evolving, which means defenders must continuously up their game just to stay competitive.

## In the Size Nines of the Cyber Attackers

When it comes to cyber attacks, the only limit is the attacker's imagination. There are no rules, no ethical boundaries. And who are the individuals carrying out the acts? Often, they are highly educated people from countries where wages are low and the standard of living is poor. Engaging in cyber crime offers them a lucrative way to escape this reality, with little immediate risk. Now, what would you do in their place? Maybe you'd make different choices – or maybe you wouldn't.

Let's move beyond ethical debates and focus on the operational side of things: How do cyber criminals achieve their goals?

## The Attacker's Arsenal

As illustrated in Figure 2.5 (drawn from IBM's 2022 "Threat Intelligence Index"), cyber criminals have a wide range of tools at their disposal. These include ransomware, server access, business email compromise (BEC), data theft, compromised login credentials, remote access, misconfigurations, malicious insiders, and other techniques such as adware, Trojan horses, botnets, cryptominers, distributed denial-of-service attacks (DDoS), worms, and more. Hackers are nothing if not imaginative.

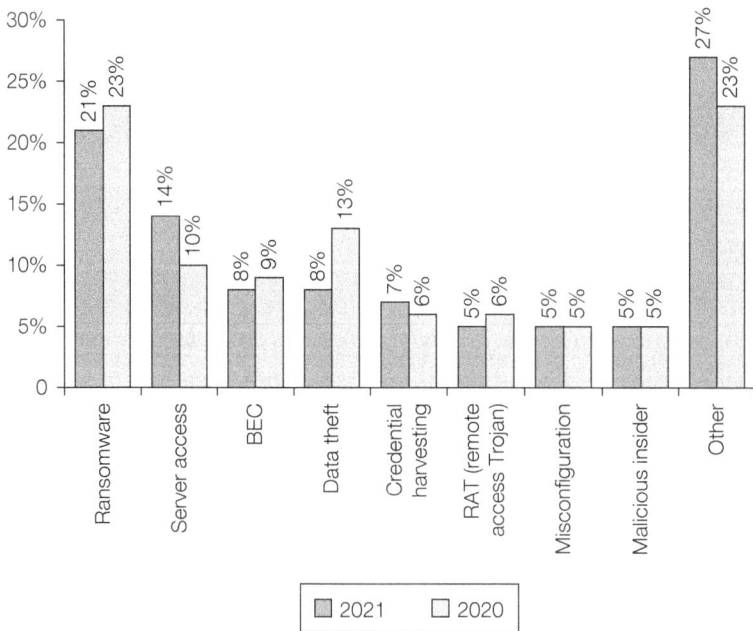

**Figure 2.5**  Evolution of the types of attackers between 2021 and 2020

*Source: IBM Security X-Force, 2022.*

The figure clearly demonstrates that the driving force behind most cyber attacks is financial gain. Ransomware, which has grown steadily since 2020, remains one of the most prevalent methods used globally.

Selected Attack Techniques:

■ **Ransomware** (see Figure 2.6 and Pocket Guide 1)

A well-known and widespread cyber-crime technique, ransomware involves sending malicious software to a victim, which then encrypts their data, demanding a ransom in exchange for the decryption key. Recent high-profile victims of ransomware include Accenture, Acer, Apple, and Colonial Pipeline.

■ **Server Access**

Attackers exploit vulnerabilities in servers to infiltrate systems and execute their attack plan.

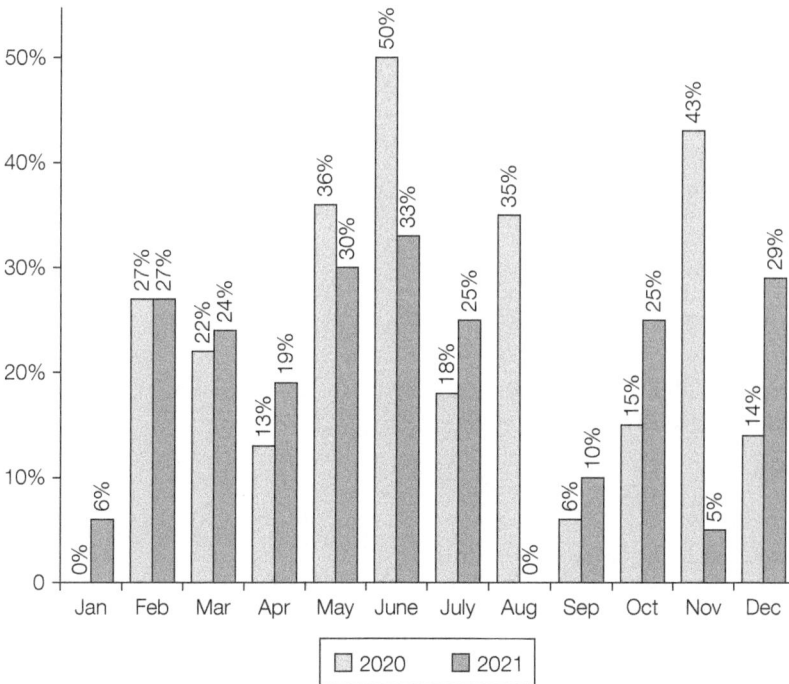

**Figure 2.6**   Incidence of ransomware events processed monthly between 2020 and 2021

*Source: IBM Security X-Force, 2022.*

■ **Data Theft**

This technique involves stealing sensitive data from company databases, devices, or servers. Notable victims of data theft in recent years include the Red Cross, the Texas Department of Insurance, the Costa Rican government, Twitter (now X), and LastPass.

■ **Business Email Compromise (BEC)** (see Pocket Guide 2)

A form of phishing attack that tricks senior executives into transferring funds or revealing sensitive information. Organisations such as Google, Facebook, the government of Puerto Rico, Toyota, and Ubiquiti have all been targeted.

■ **Compromised Identifiers**

Also known as credential harvesting, this method targets user credentials (such as usernames, passwords, and email addresses) to gain unauthorised access to systems.

■ **Remote Access**

Using malware disguised as legitimate software, attackers gain full administrative privileges and control of a target system.

■ **Using a Configuration Fault**

Misconfigurations can leave applications or entire systems exposed to attacks. Organisations affected by such flaws include Amazon, Citrix, and NASA.

■ **Malicious Internal Actor** (see Pocket Guide 3)

Sometimes called a malevolent insider, this type of attack involves a person within the organisation, such as a current or former employee, who intentionally misuses proprietary information. Bupa, Cisco, and General Electric have all been affected by such breaches.

■ **Adware**

This software displays unwanted advertisements or downloads them onto devices during program operation.

■ **Trojan Horse**

Trojan-horse malware is hidden within seemingly legitimate software. Once installed, it grants the attacker backdoor access, allowing for espionage or data theft. Recent victims include Amazon, Bank of America, and Cisco.

■ **Botnet**

A botnet consists of a network of malware-infected computers under the control of a single entity, often used for large-scale attacks.

■ **Cryptominer**

Malware that hijacks a system's processing power to mine cryptocurrencies. Google Cloud reported that 86% of compromised cloud instances were used for cryptomining in 2021.

■ **Website Degradation**

Also known as defacement, this attack damages a website's functionality or content. Victims of such attacks include Ashley Madison, the US government, the Indian Ministry of Defence, and Bitcointalk.org.

■ **President Fraud** (see Pocket Guide 2)

Hackers impersonate senior executives and direct employees to make fraudulent financial transfers. This is a type of BEC attack.

■ **Distributed Denial of Service (DDoS)**

A DDoS attack aims to shut down a system by overwhelming it with traffic. Victims include Amazon, OVH, and JPMorgan Chase.

■ **Worms**

Worms are malicious programs that self-replicate and spread across networks, overwhelming systems with traffic.

■ **Malicious Scripts**

Modified code designed to compromise web applications or websites.

■ **Waterhole**

Cyber criminals infect websites that are frequently visited by members of a target group, hoping to compromise one of their devices when they visit the infected site.

■ **Typosquatting**

This technique targets users who accidentally mistype web addresses. Equifax and YouTube have both been affected by this attack.

■ **Phishing**

One of the most common types of social engineering attacks, phishing uses fake communications to trick individuals into revealing sensitive information or clicking on malicious links. Crelan Bank, Facebook, and Google have all been victims of phishing attacks.

## The Main Infection Vectors

Now that we've explored the techniques cyber criminals use to infiltrate information systems, let's look at their targets. According to IBM's 2022 "Threat Intelligence Index", two clear trends emerge: machines and humans. While defenders primarily rely on technology to secure systems, attackers also target employees, often exploiting human weaknesses just as much as technological vulnerabilities.

Figure 2.7 illustrates how the infection vectors are split: roughly 50% technical and 50% human. One of the most significant vulnerabilities in any information system is often found, as the saying goes, somewhere between the screen and the chair. When you consider that defenders need to block every attack, while attackers only need to succeed once, the necessity for layered security becomes clear. A well-structured defence should follow the PACE model (see also Chapter 6) – Primary, Alternate, Contingency, and Emergency – and grow in sophistication as the organisation evolves.

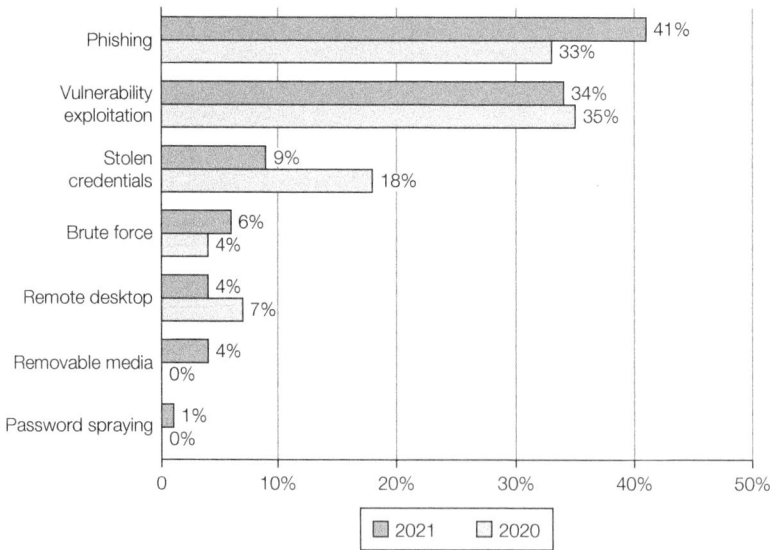

**Figure 2.7**  **Main vectors of infection between 2020 and 2021**

*Source: IBM Security X-Force, 2022.*

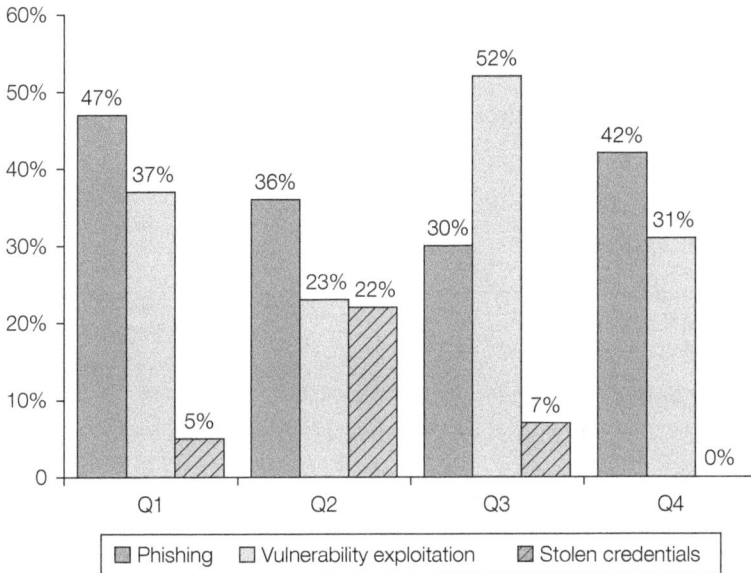

Source: IBM Security X-Force, 2022.

**Figure 2.8**  **Attack targets by quarter in 2021**

## Between the Screen and the Chair

Interestingly, Figure 2.8 reveals a seasonal trend in attack strategies. Cyber criminals are opportunistic, and they time their attacks to maximise impact. The data shows that human-targeted attacks peak during two significant periods: global tax deadlines. On the other hand, technical vulnerabilities are more often exploited during times like the summer holiday season when fewer employees are actively monitoring systems. Attackers always find the right time to strike, leveraging human distraction or system weaknesses whenever possible.

# Anatomy of a Cyber Attack

A cyber attack unfolds in six stages, grouped into four main sequences (as shown in Figure 2.9). Understanding these stages helps defenders anticipate the tactics used by cyber criminals and develop a comprehensive defence strategy.

1  Infiltration: The attacker uses an entry point, like a phishing email, an infected file, a cloud misconfiguration, or a vulnerability in an application, to introduce malicious software into the system.

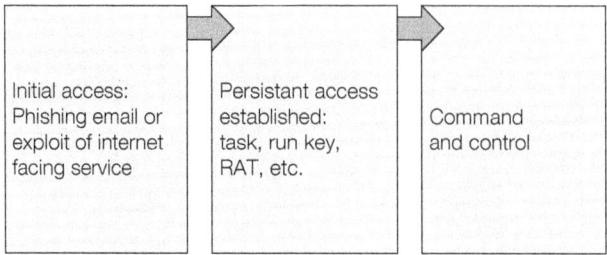

**Figure 2.9**  **Anatomy of a cyberattack (sequence 1)**

*Source: IBM Security X-Force, 2022.*

2   Reconnaissance: The malicious software scans the network for further vulnerabilities and potential access points. During this phase, the malware communicates with the attacker's external servers to receive instructions or additional code.

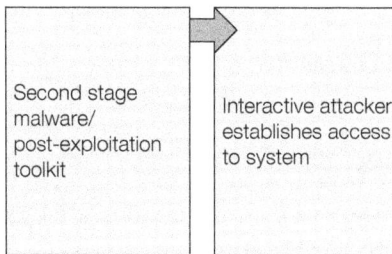

**Figure 2.9**  **Anatomy of a cyberattack (sequence 2)**

*Source: IBM Security X-Force, 2022.*

3   Establishing Persistence: The hacker creates new backdoors to maintain access, even if some of their entry points are discovered and closed.

4   Lateral Movement and Data Collection: Once inside, the attacker gathers login credentials and uses them to move deeper into the network, gaining access to sensitive systems and data.

| Gather credentials:<br>• Local reconnaissance:<br>Local users, groups, tasklist, etc.<br>• Active Directory reconnaisance:<br>Gather list of domain admins and domain controllers | SMB lateral movement | Reconnaissance | Gather credentials:<br>• Gather lists of hostnames/subnets to target for ransomware.<br>• Gather lists of data sources |
|---|---|---|---|

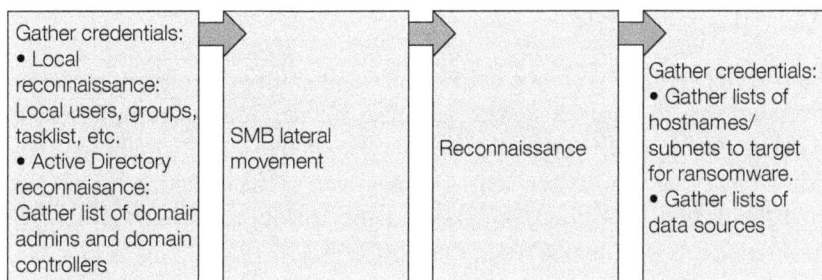

**Figure 2.9** Anatomy of a cyberattack (sequence 3)

*Source: IBM Security X-Force, 2022.*

5  Exfiltration: The stolen data is stored on a bounce server, ready for extraction. At this point, sensitive data and systems are primed for encryption or exfiltration.

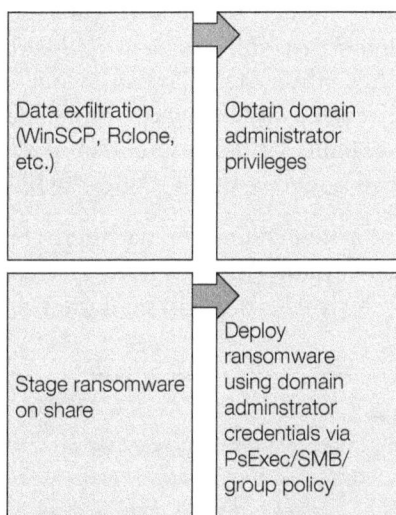

| Data exfiltration (WinSCP, Rclone, etc.) | Obtain domain administrator privileges |
|---|---|
| Stage ransomware on share | Deploy ransomware using domain adminstrator credentials via PsExec/SMB/group policy |

**Figure 2.9** Anatomy of a cyberattack (sequence 4)

*Source: IBM Security X-Force, 2022.*

6  Ransom or Sale: After exfiltration, the data is either sold on the darknet or used to demand a ransom from the victim to unlock encrypted data or regain control of compromised systems.

> *"The best time for the attacker will always be the worst time for the victim."*

Attackers are patient and methodical. They will wait until the perfect moment to strike when defenders are least prepared, creating maximum chaos – a key lesson when it comes to preparing for cyber crisis scenarios.

# The Weakest Link

We've touched on this earlier, but it's worth reiterating. Companies need to invest not only in technical defences but also in staff training. The importance of human factors in cyber security was demonstrated in a 2010 experiment carried out by cyber security specialist Thomas Ryan in the United States, which we'll discuss in detail below. The case highlights that humans, not technology, are often the most vulnerable part of an information system. This vulnerability explains why cyber criminals focus more than half of their access attempts on targeting people, rather than technology alone.

Before we dive into the specifics, we'd like to share a conversation we once had with a French fighter pilot. Interestingly, the digital world isn't the only place where human limitations are the weakest link. The pilot was proudly boasting about the capabilities of his aircraft, and, like any normal adrenaline junkie would, we listened attentively. Our conversation eventually shifted to combat drones. The pilot mentioned that while his plane could accelerate faster and turn more sharply than the control system allowed, he couldn't use those maximum capabilities. When we asked why, he replied with a smile, "Despite my suit, if I pushed the aircraft to its limits, I'd likely pass out." And without a pilot at the controls, a 10-tonne aircraft would eventually drop out of the sky like a stone. . .

This is a reminder that no system, human or machine, is bulletproof forever. In cyber security, just as in aviation, there is no such thing as zero risk. The human factor is, and always will be, the most significant variable.

---

**Case Study: Robin Sage**

Meet Robin Sage, a plucky young cyber security expert based at the Naval Network Warfare Command in Norfolk, Connecticut. According to her LinkedIn profile, Robin is a 25-year-old graduate of the prestigious MIT and one of the few women in her

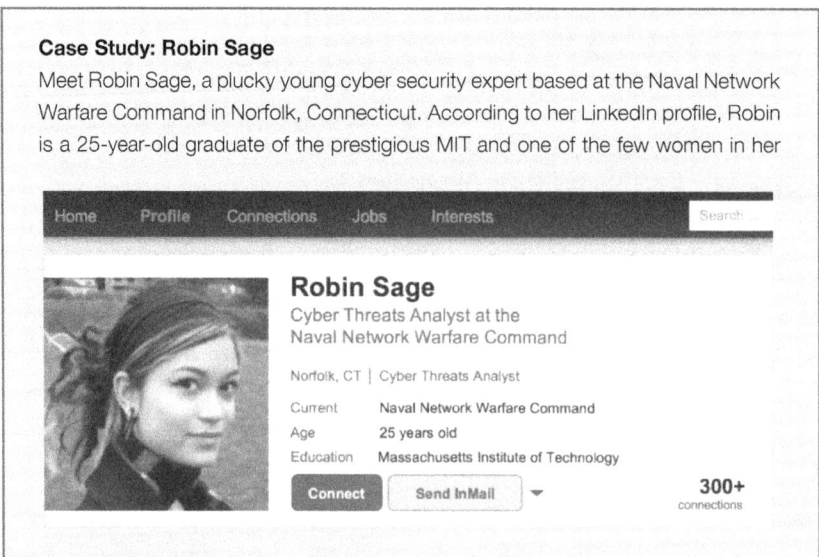

| Home | Profile | Connections | Jobs | Interests | | Search |
|------|---------|-------------|------|-----------|--|--------|

**Robin Sage**
Cyber Threats Analyst at the
Naval Network Warfare Command

Norfolk, CT | Cyber Threats Analyst

| | |
|---|---|
| Current | Naval Network Warfare Command |
| Age | 25 years old |
| Education | Massachusetts Institute of Technology |

Connect | Send InMail ▼

**300+**
connections

field. She's already built an impressive network of over 300 industry contacts. But there's one small detail: Robin Sage doesn't exist.

The Robin Sage experiment was created by cyber security researcher Thomas Ryan in 2010 to demonstrate the vulnerabilities in professional networking and the risks posed by social engineering. For 28 days, Ryan monitored the fake profile, named after an unconventional warfare exercise used by the Kennedy Special Warfare Center. The profile photo was that of an actress, unrelated to the military or cyber security.

Despite being a fabricated persona, Robin quickly connected with hundreds of industry professionals, including individuals from the Department of Defense, NSA, defence contractors like Lockheed Martin and Northrop Grumman, and tech companies like Google. During the experiment, Robin was even provided with sensitive military helicopter take-off times by a soldier on active duty and received multiple job offers from reputable organisations.

Ryan presented the results of this experiment at the Black Hat conference in 2010. The outcome highlighted how easily a well-crafted social engineering profile can deceive even seasoned professionals. Even those with a robust understanding of security – both physical and digital – fell into the trap set by Robin's carefully constructed online presence.

The Robin Sage experiment serves as a potent reminder of the importance of vigilance in cyber security. If security-conscious professionals from the world's top defence and technology agencies can be tricked by a LinkedIn profile, then no one is entirely immune to the risks of social engineering.

### Key Takeaways

In Chapter 2, we've shown that cyber crime is no longer just a loose network of opportunistic individuals but rather an organised, well-financed industry with codes, rules, and a cast of players who move with purpose. Far from the image of a lone teenager in a hoodie, modern cyber criminals operate in teams, often with substantial resources at their disposal, and in some cases, even a fleet of luxury cars.

Before we go into the next chapter – where we'll see how IT and cyber security teams can proactively prepare for cyber threats – let's take a moment to summarise the key insights we've covered.

▶

**Cyber Crisis Management:**

- **A Matter of When, Not If**

  Every organisation, regardless of size or industry, is at risk. Preparing for the inevitable is now a fundamental part of managing cyber security risk.

- **Cyber Attacks Are a Constant Threat**

  From public bodies to private companies, the headlines reflect a weekly litany of major incidents – each one a stark reminder of the pervasive nature of cyber crime.

- **Profit Is the Primary Driver of Cyber Crime**

  The majority of cyber attacks are fuelled by the lure of financial gain. To maximise their return on investment, cyber criminals evolve, specialise, and even collaborate within a guerrilla-style ecosystem of digital threats.

- **Attacks Are Growing in Complexity**

  Attackers only need to find a single vulnerability, while defenders must ensure their defences are comprehensive and up to date at all times. In this asymmetrical battle, the odds are stacked in favour of attackers.

- **Human and Technical Vulnerabilities Are Prime Targets**

  Cyber criminals exploit a combination of technical flaws and human error to infiltrate, navigate, and seize control of systems. Focusing exclusively on technical measures ignores over half of the initial attack vectors, leaving organisations exposed to human-targeted attacks.

  These points underscore the strategic thinking that must underpin any successful cyber security effort. The evolving landscape of cyber crime demands a layered, adaptive approach to defence that incorporates not just technology, but people and processes too.

# 3

# PREPARING YOUR IT TEAMS

"No executive cares about a security strategy; what they want
is a secure business strategy."

– Shamla Naidoo

All professionals – from athletes to artists, firefighters to surgeons – agree on a universal truth: if you want to be a master of your art, you need to keep honing your skills. The cyber-threat landscape is no different. Staying sharp in this field means committing to ongoing training and keeping pace with new developments. As Henry Ford famously put it, "The only thing worse than training your employees and having them leave is not training them and having them stay."

A stark reminder of the necessity for such preparation came in 2020 when major software player SolarWinds became the target of one of the largest and most sophisticated cyber attacks in history. The attackers exploited a vulnerability in the company's software updates, which ultimately granted them access to the networks of thousands of other organisations, including several US government agencies. This incident not only compromised sensitive data but also revealed critical vulnerabilities in the approach many organisations take towards security.

The severity of the SolarWinds breach underscores the importance of proactive internal testing and training for IT teams. Had SolarWinds invested in regular simulated-attack scenarios and comprehensive training for its staff, the firm might have detected the malicious activity sooner. By conducting thorough penetration tests and fostering a culture of continuous learning, SolarWinds could have enhanced its security posture and response capabilities, potentially mitigating the impact of the breach.

Leading companies like Microsoft and Google exemplify how regular training and exercises can fortify defences against sophisticated threats. By running internal simulations that mimic real-world cyber incidents, these organisations enhance their detection capabilities while building team cohesion and readiness. This proactive approach not only prepares teams for the unexpected but also reinforces the critical message that security is an ongoing effort, not just a box to tick.

The SolarWinds Hack: The Largest Cyber Espionage Attack in the United States.

In this relentless battle against cyber threats, it's essential to equip IT teams with the skills, knowledge, and real-world experience needed to stay one step ahead.

# IT Security: The Core of Your Cyber Defence

At the heart of any effective cyber-defence system is the IT security team. The people in these teams lead the charge when crises strike and form the backbone of an organisation's cyber resilience. As Thomas Billaut, Head of Cyber Operations at Forvia, wisely points out, "Cyber teams that don't stay up to date on the latest techniques are like unsharpened knives – they eventually forget what they're designed for."

What sets these teams apart isn't just their technical prowess; it's their ability to adapt in real time, anticipate potential threats, and mobilise an effective response that safeguards the organisation's assets and reputation. Think of them as the architects of a digital fortress, constantly redesigning their defences to thwart the latest siege tactics employed by attackers.

For instance, in 2017, Equifax experienced one of the most significant data breaches in history, impacting 147 million customers. The breach occurred primarily because a known vulnerability wasn't patched. However, the incident also showcased the importance of collaboration between IT security and other departments. Following the breach, Equifax improved its incident response by encouraging its IT security team and other business units to work together much more closely. By improving the flow of communication and conducting regular joint training sessions, they established a culture of vigilance and preparedness that has since helped the company mitigate future threats more effectively.

As threats become more complex and more frequent, collaboration is not just beneficial – it's essential for a unified and agile defence. The Chief Information Officer (CIO) manages the operational framework, ensuring business processes run smoothly, while the Chief Information Security Officer (CISO) focuses on shielding those processes from digital threats. A notable example of this partnership can be seen in the proactive measures taken by the Marriott

International CIO and CISO following a data breach in 2018. By working closely together, they implemented enhanced security protocols, including improved encryption methods and more robust access controls. Their collaboration not only strengthened Marriott's defences but also set a precedent for cross-departmental teamwork in cyber security risk management.

When the CIO and CISO align strategically, the result is more than just clear priorities; it's a streamlined approach that reduces risks, mitigates internal conflict, and ensures that security is a foundational element of the business strategy, not just an afterthought.

## Know Your. . . Ally

Cyber Threat Intelligence (CTI) is the silent sentinel, quietly steering cyber-defence strategies in sync with the ever-evolving threat landscape. But CTI isn't just about amassing data; it dives deep into analysis, interpretation, and pattern recognition. This allows companies to foresee potential attacks and fortify their defences ahead of time. Think of CTI as your organisation's very own "spy network", empowering defenders with the insights they need to stay one step ahead. Unlike attackers, who only need to succeed once, defenders must maintain vigilance in the face of every single threat.

Your allies in this ongoing battle include government agencies, private intelligence consultancies, and dedicated in-house CTI teams. Their insight helps you form a robust cyber strategy, integrating real-time threat detection, continuous monitoring, and coordinated response plans. As Sun-Tzu wisely noted, "He who has no objectives is unlikely to achieve them." In the realm of cyber security, CTI lays down those crucial objectives, steering companies away from reactive measures and towards a proactive strategy.

It is essential that CTI translates into actionable insights for front-line teams; organisations must prioritise the accessibility and relevance of intelligence. Strong cyber security leadership integrates CTI insights into the daily routines of operational teams, ensuring that intelligence informs decisions and shapes long-term planning. For example, organisations like IBM have successfully incorporated CTI into their cyber security frameworks, leading to sturdier defences and quicker response times to threats.

When executed correctly, CTI enables organisations to predict, adapt, and respond with agility, cultivating a culture of resilience that empowers both individuals and teams. Cyber defence is not a static shield; it's a dynamic strategy where intelligence continuously sharpens response capabilities, transforming how we tackle the complexities of modern threats.

## Leveraging Intelligent Automation

In elite fields – whether sports, military operations, or emergency response – the synergy between "intelligence" and "operations" is paramount. The military sums this up nicely with the phrase, "intelligence informs operations." It's a principle that resonates across disciplines, but cyber security presents its own unique challenges. Unlike many other sectors, cyber security professionals must defend against threats that never abate, and leave little room for pre-game warm-ups or controlled drills. For them, the game is perpetually in play, and this constant pressure underscores the critical importance of intelligent automation.

Statistics highlight the impact of automation on cyber security effectiveness. A recent study by the Ponemon Institute[1] found that organisations using intelligent automation in their security operations reported a 36% improvement in their ability to detect and respond to security incidents. This demonstrates the need for companies to embrace automation as a key component of their security strategies.

Intelligent automation serves as a lifeline for cyber teams, providing them with much-needed breathing room to reassess and refine their strategies rather than just giving them the tools to put out fires. In fact, research shows that organisations leveraging intelligent automation can reduce incident response times by up to 60% and improve overall threat detection accuracy by 75%. However, the adoption of automation in cyber security hasn't yet reached its potential. A significant barrier is the misconception surrounding its implementation – nearly 40% of companies express concerns that automation might inadvertently block legitimate activities or fail to catch genuine threats. This scepticism, compounded by the time constraints that security teams face, presents a paradox with which many organisations continue to grapple.

Let's address some common misconceptions. Many professionals worry that automation could lead to job losses or decrease the necessity for skilled personnel. However, the reality is that automation is designed to enhance human capabilities rather than replace them. For example, a case study on IBM Security shows that organisations using automation saw a 30% increase in their security teams' efficiency, allowing them to focus on more complex threats that require human insight and creativity.

Another example comes from JPMorgan Chase, where automation was implemented to enhance the company's fraud-detection capabilities. Initially

---

[1]   "State of AI in cybersecurity", 2024 Report, Ponemon Institute.

hesitant due to fears of missing genuine transactions, the bank discovered that automation actually improved accuracy, allowing teams to focus on more complex cases while the automated system efficiently handled routine alerts. This real-world application dispels the myth that automation diminishes oversight; instead, it enhances overall efficacy.

When properly implemented, automation can handle routine tasks, freeing up cyber professionals to engage in strategic planning and critical thinking. This is where artificial intelligence (AI) shines, enhancing human capabilities and allowing teams to focus on complex, adaptive tasks. Automation has already shown its excellent ability to manage rule-based, repetitive processes, but cyber security is no "set-and-forget" endeavour. The threat landscape is constantly shifting, demanding that our defences evolve with it. As we highlighted in Chapter 2, automation in cyber security must be intelligent, dynamic, and adaptable; it must keep pace with emerging threats.

> *"Leaving the monopoly on innovation to the attackers is like handing someone a knife to stick in your back."*

> – Yann Bonnet

A practical entry point for automation is focusing on known, repeated threats. While this may seem quite basic and even obvious, in the world of cyber security simplicity brings the best results. As AI engines continue to advance, they will increasingly handle more sophisticated tasks, liberating human teams to confront challenges that require creativity and critical thinking. Rather than replacing humans, automation complements the existing framework of intelligence and operations, enabling people to focus on high-impact threats instead of getting bogged down by lower-level distractions. Intelligent automation doesn't just enhance operational capacity, it empowers cyber security teams to navigate an increasingly complex and dynamic landscape with confidence.

Evaluate your existing processes – where are the bottlenecks? Which repetitive tasks could be automated to free up your teams for more critical work? By identifying opportunities for integration, companies can enhance efficiency and responsiveness, ultimately strengthening their cyber-defence capabilities.

## Staying in the Race

Now that we've freed up all that time and essential resources (easy, wasn't it?), let's move on to the critical area of operational training. Imagine a high-stakes training scenario where cyber security professionals are thrust into a simulated crisis. Everyone goes into high alert when they receive news of an incident: a ransomware attack has compromised critical systems. With only a few hours to respond, the team must spring into action and analyse the situation, identify vulnerabilities, and coordinate their response. It's a race against time.

And like elite forces, cyber security teams thrive on realistic, high-pressure training and a strong team spirit. They need expert instructors who can guide them up to their limits, challenging them in ways that build both skill and resilience, since real attackers won't show any mercy. Instructors who excel in delivering this type of training must create environments that accurately reflect real-world systems, vulnerabilities, and sector-specific contexts. The goal is to be as realistic as possible. Once the environment is set up, the technical teams log into an online platform or visit a training centre, detaching from their daily routines to immerse themselves in a controlled yet intense scenario:

- ■ If operational teams work across multiple time zones, virtual training may be the best fit.

- ■ When teams are used to collaborating closely in person, a physical training centre can reinforce team cohesion and provide a more immersive experience.

*"As the saying goes: train like you fight."*

Rotating training sessions every six to twelve months is ideal, enabling each team member to benefit and refine their skills over time. These sessions are best held as intense, day-long drills. From start to finish, team members juggle various incidents, conduct technical investigations, coordinate with IT, and produce concise reports for cyber security managers. In turn, these managers deliver executive summaries to senior leadership, guiding critical decisions beyond the technical realm (a concept we'll explore further in Chapter 5).

This collective approach not only enhances operational efficiency but also cultivates a strong team dynamic. When participants engage in realistic scenarios together, they build trust and camaraderie that are vital during real crises. For example, after undergoing joint training exercises, a financial institution's incident response team improved their communication during actual incidents, leading to quicker resolutions and minimised downtime.

As Frank Van Caenegem, Cybersecurity Vice-President and CISO EMEA at Schneider Electric, says, "It is just as important to present the deputies of CISOs to management committees as it is to present the CISOs themselves. This enables deputies to develop cross-functional skills like leadership and communication and to gain managers' trust if backup is required."

After each session, it's essential to request an instructor's report. Observing both individual and group performance helps identify progress and areas for improvement. It also demonstrates the value of your time and investment in training. Evaluating how effectively your teams respond to real incidents after their training is another crucial measure of success.

Training isn't a one-time event; it's an ongoing journey. Think of it like a marathon: preparation is crucial, but the learning continues even after the finish line. Companies like FireEye, which conduct regular simulated-attack scenarios, have seen improved incident-response times and a heightened state of readiness across their teams.

Experienced instructors will push expert teams to excel with training that is realistic, grounded in current threat scenarios, and well structured, ensuring that teams are prepared when it counts most. Preparing one-day sessions at this level of intensity can take instructors up to a hundred hours, but the investment pays dividends in the real-world resilience it builds.

## Mission Alignment

We understand the importance of keeping cyber security experts on the cutting edge, but what about the broader Information Systems Management (ISM) teams? It's important to bring ISM into the fold because they handle the operational backbone – installing security patches, managing data, and developing applications. Without their involvement, digital risk mitigation becomes a much steeper uphill battle.

Consider the notorious 2017 Equifax data breach, in which attackers gained access to the personal information of approximately 147 million people. One significant factor was the failure to patch a known vulnerability in a timely manner. If the ISM personnel had been better trained to prioritise security during software updates and understood the implications of their actions on overall security posture, the breach could perhaps have been mitigated. This incident demonstrates the necessity of having ISM teams that are not just operational but also highly conscious about security.

The pathway is clear: aligning ISM training with cyber security objectives ensures that these teams are not only aware of their responsibilities but also equipped with the skills needed to address potential risks. When ISM personnel understand how their actions contribute to the broader security framework, they become proactive allies in the fight against cyber threats.

By integrating regular training sessions that focus on both technical and security aspects, organisations can create a culture where security is woven into the fabric of daily operations. This proactive stance is vital, as it empowers ISM teams to identify potential threats early and act before vulnerabilities can be exploited.

To further illustrate the importance of mission alignment and collaboration between these teams, consider the partnership between Marriott International's Chief Information Officer (CIO) and Chief Information Security Officer (CISO). Following a significant data breach in 2018 that exposed the personal information of millions, their combined efforts led to an overhaul of Marriott's cyber security measures. This partnership fostered improved communication channels, which enabled speedier responses during incidents and established protocols for integrating security into all aspects of business operations. By aligning their goals, the CIO and CISO not only enhanced the company's security posture but also cultivated a culture of cyber security awareness.

As threats increase in number and become more complex, collaboration between IT and cyber security teams is not just beneficial, it's essential for a unified and agile defence. The CIO manages the operational framework, ensuring business processes run smoothly, while the CISO focuses on shielding those processes from digital threats. When the CIO and CISO align strategically, they don't just generate clear priorities; they create a streamlined approach that reduces risks, mitigates internal conflicts, and ensures that security is a cornerstone in the business strategy.

## Backup and Running

In a cyber crisis, controlled, secure backups are paramount. Techniques like immutable backups are critical, but for these to be effective, backups must be properly configured and isolated from possible compromises. If a breach slips past initial defences, a reliable backup may be the only solution. But here's the rub: attackers know this and often aim to compromise backups, expecting that companies will restore from them and give the attacker more leverage.

This interdependency reinforces that security is a team effort across IT, IS, and management. Digital assets are not effectively protected with strong tech alone, but also require active collaboration across departments, making every team integral to a firm's cyber resilience.

In recent years, organisations in many different industries have recognised the need to embed cyber security directly into their strategic business initiatives. Many leading firms are now establishing cross-functional teams that blend IT and cyber security expertise to proactively assess and fortify cyber defences across critical projects. These teams continuously analyse digital risks, recommend protective measures, and ensure that cyber security considerations are integral to each stage of their digital transformation.

This proactive model, which focuses on early involvement and thorough analysis, enhances awareness about cyber security across departments and establishes protective measures as foundational elements rather than last-minute additions. Over time, this integrated approach contributes to a culture of security throughout the organisation, which fosters a self-sustaining cycle where cyber security knowledge is transferred widely, and digital resilience becomes a core element in business operations. The ultimate goal is to create an environment where dedicated cyber security interventions are required less frequently, which reflects a mature, resilient security culture.

## Positive Internal Attacks

Identifying and rectifying your own vulnerabilities before external attackers can exploit them is a crucial strategy for any organisation. Many are now recognising the value of proactive internal testing, yet committing to such efforts requires both courage and determination. A notable example comes from a prominent English-speaking company where the head of cyber security methodically expanded her penetration-testing budget. This approach prioritised internal testing as an essential part of operational security.

She instructed her team of "ethical hackers" to reach into the company's systems and extract any vulnerabilities they could find – no holds barred. Initially, this strategy met resistance from divisional managers concerned about potential disruptions to production or delays in software rollouts. However, over time, her leadership transformed the organisation's approach to cyber security, fostering a culture where security was viewed as an ongoing, dynamic process rather than just a compliance checkbox.

The analogy of vaccines serves as a powerful tool to explain how controlled exposure can strengthen defences. Just as vaccines introduce a small, harmless dose of a virus to build immunity, organisations can simulate cyber threats to prepare their defences. By conducting internal penetration tests, employees are trained to recognise vulnerabilities and respond effectively to potential attacks. This proactive mindset enhances the company's own resilience and ensures that teams are not only aware of the risks but are also equipped with the knowledge to mitigate them.

**Case Study: American Express**

Consider the case of American Express, which has successfully integrated internal penetration testing into its security framework. By regularly conducting internal simulations, the company empowers its cyber security team to identify weaknesses before they can be exploited. This practice has yielded significant benefits:

■ **Enhanced Detection Capabilities**
American Express found that regular penetration testing allowed them to detect vulnerabilities earlier, minimising the risk of breaches.

■ **Increased Employee Awareness**
The testing not only improved security protocols but also heightened employee awareness of cyber security issues throughout the company, creating a more vigilant workforce.

■ **Stronger Team Dynamics**
The process fosters collaboration between IT and security teams, reinforcing the notion that cyber security is a collective responsibility.

Through these internal tests, American Express has created an environment where proactive measures are part of the company culture, underscoring the importance of viewing cyber security as an ongoing commitment rather than a one-time task.

# How to Conduct and Evaluate Training Sessions

To keep cyber security teams and IT departments sharp, realistic training is essential – and so is evaluation. Organisations today must move beyond merely conducting training sessions; they need to measure their effectiveness rigorously. For instance, companies like Cisco report a 30% reduction in incident response times after implementing regular simulations. Such statistics underscore the value of ongoing training and preparation.

## Measuring Performance

Let's start with cyber security teams, who are often already skilled in dealing with incidents, threats, and attacks. Keeping their knowledge up to date is crucial since cyber defence is a relentless game of catch-up against evolving threats. When it comes to training, instructors need to get inside the heads of potential attackers, simulating how they would find vulnerabilities within the company's systems.

A notable example is IBM's X-Force team, whose members train their internal teams using simulated attacks based on real-world incidents, such as ransomware and data theft. By focusing on specific vulnerabilities identified through threat intelligence, IBM tailors each session to stress-test systems and sharpen response tactics. Through this method, companies can mirror realistic scenarios to drive home the urgency of proactive defences.

## Scenario Design

Effective training scenarios should push teams to their limits, but means the instructor must have a thorough understanding of the organisation's infrastructure and common entry points for attackers. Before the session, instructors should interview IT and cyber teams to design scenarios that feel genuine. For example, suppose a company recently saw an increase in phishing attempts. In that case, a training session might involve simulated phishing attacks where team members must identify and mitigate these attempts under pressure, reflecting the exact conditions they're likely to face.

## Evaluating Training Impact

After training, assessment is just as important. Beyond ticking boxes, evaluation metrics should focus on measurable improvements, such as response times, accuracy in threat detection, and teamwork under stress. Many companies now employ "after-action" reviews, where teams reflect on what worked and what didn't, identifying areas for improvement. For example, Verizon reported a 20% improvement in their security posture following comprehensive post-training evaluations that informed future training sessions.

In the end, well-structured training not only enhances technical skills but also boosts team confidence and agility – ensuring that when a real threat comes, the organisation is ready.

## Bridging Theory and Practice

Once the foundational knowledge has been established, the actual training phase begins, and that's a critical component in which theory meets practice. It's where the rubber hits the road. Participants are granted access to a simulation platform that will become their battlefield for the hours ahead.

As the clock starts ticking, they dive into approximately half a day of intense, hands-on engagement. They are thrust into a whirlwind of activity that includes forming hypotheses, conducting investigations, and collaborating with teammates. The atmosphere is charged with urgency as they prioritise corrective actions, isolate threats, and implement countermeasures – all while navigating the complexities of a simulated cyber crisis.

Instructors play a crucial role in this process, maintaining a firm but supportive presence. They resist the temptation to reveal any insights or shortcuts, understanding that creating a false sense of security can be just as detrimental as inadequate preparation. This commitment to realism allows participants to confront the challenges head-on, mirroring the unpredictable nature of actual cyber threats.

One effective method to enhance this training is through war-gaming exercises. These simulations allow participants to engage in scenarios that replicate real-world attacks, such as ransomware threats or phishing attempts. For instance, during a war game conducted for a healthcare provider, participants faced a simulated ransomware attack on the organisation's patient management system. They had to make rapid decisions about data recovery and communication strategies with no hints or guidance from the instructors. This approach tested their skills under pressure, fostering an environment where learning and growth can flourish.

In some cases, the sessions span several days and involve hundreds of participants, amplifying the complexity and realism of the training experience. Such lengthy exercises often take place on company premises, which makes the experience more immersive and authentic. However, this heightened level of engagement comes at a cost – both in terms of resources and financial investment. Companies must commit significant time and money to prepare for these large-scale simulations, but the benefits can be transformative.

Organisations like Northrop Grumman and Cisco use war gaming as part of their training regimen. Northrop Grumman conducts tabletop exercises that simulate various cyber-attack scenarios, improving communication and collaboration during real incidents. During Cisco's cyber-defence exercises, cross-functional teams develop unified response strategies and their security posture is continually assessed.

These extended training sessions not only challenge participants but also promote cross-departmental collaboration. With more individuals involved, teams from various functions – such as IT, operations, and legal – can work together to address the multifaceted nature of cyber threats. This collective approach fosters a deeper understanding of how their roles interconnect, ultimately strengthening the organisation's overall resilience.

The training phase is where participants begin to internalise the critical lessons of cyber crisis management. By the end of these sessions, they emerge not just as individuals equipped with knowledge but as cohesive teams ready to tackle the evolving landscape of cyber threats head-on.

---

**Top Tip – Into the Wild. . .**

If you decide to conduct your training in a real-world environment, it's essential to establish a robust communication loop with your internal crisis management unit. This proactive measure ensures that everyone is on the same page and minimises confusion during the exercise.

- ■ Start by disseminating a notice well in advance, sharing the contact details of the internal crisis unit with as many employees as possible. This way, everyone knows who to contact in case of an emergency, whether it's part of the training or an actual incident. Clear communication channels are vital, especially in high-stress scenarios requiring rapid responses.
- ■ It might also be wise to inform national cyber-defence units about the training exercise ahead of time. Alerting them to the possibility of false alerts not only fosters collaboration but also prevents unnecessary panic and avoids the drill being misinterpreted as a real cyber incident. For instance, during a recent training exercise conducted by a UK financial institution, officials coordinated with the NCSC (National Cyber Security Centre) to ensure that the simulation was recognised as an exercise, avoiding any disruptions to their operations or response protocols.
- ■ When these communication protocols are established, the training can be a more realistic and effective experience. It ensures that participants immerse themselves fully in the exercise without the distraction of external alarm or confusion, and the exercise is a true test of their skills and readiness.

By reinforcing the importance of communication in your training exercises, you not only prepare your teams for the challenges they may face during a cyber crisis but you also build a culture of awareness and preparedness throughout your organisation.

---

## Evaluation

After each training session, transitioning into a comprehensive review phase is crucial to maximise learning outcomes. This phase typically begins with a detailed observation report that highlights both the strengths demonstrated during the exercise and the areas that need improvement. Such reports are invaluable for shaping future training courses because they enable teams to track their performance over time and adapt training content to better meet their evolving needs.

To ensure meaningful progress, consider the following actionable insights:

■ **Comparison of Consecutive Reports**
Analyse performance trends by comparing reports from previous training sessions. This allows you to identify specific areas of improvement and measure performance gains.

■ **Targeted Interventions**
If a team consistently struggles with a particular aspect of incident response, design targeted training sessions focused on that gap. For example, if people are struggling with communication during crises, organise workshops that simulate communication scenarios.

■ **Recognition of Team Resilience**
Acknowledge that increased difficulty doesn't always correlate with improved performance for every individual. Focus on the team's collective ability to face challenges and grow together. Highlight instances of resilience to encourage a growth mindset.

■ **Regular Rotations Among Sub-Groups**
Conduct simulations with different sub-groups within the cyber security teams to expose various members to diverse scenarios. This practice helps foster collaboration and shared learning.

■ **Continuous Refinement of Training Programmes**
Use insights gained from evaluations to refine training programmes continuously. This ensures that training remains relevant, effective, and aligned with the organisation's evolving needs.

Thorough evaluations after each training exercise provide critical insights that inform future sessions. By implementing these actionable points and ensuring regular participation across various sub-groups, organisations can continuously enhance their training programmes, ultimately building a more agile and prepared cyber security workforce.

## Simulating System Restarts

Cyber defence is inherently a team sport, requiring each participant to fulfil their role effectively to minimise damage during an incident. This makes IT-team training a vital part of a company's overall cyber strategy. The responsibilities of IT teams extend beyond simply maintaining smooth operations; they are the frontline responders during a digital crisis, swiftly rebuilding servers and executing backup protocols to restore normalcy amidst chaos.

The complexity of modern information systems can be overwhelming, often likened to a plate of spaghetti. They are made up of layers and interconnections that obscure clear boundaries and decision-making paths. As organisations demand greater agility, they sometimes fail to equip IT teams with the necessary resources to meet these expectations. For instance, isolating a system to prevent malware spread might sound straightforward, but executing that plan can be fraught with complications.

So, how do you prepare IT teams – primarily tasked with ensuring digital infrastructure stability – to tackle these challenges effectively? The answer lies in simulation. Training can take place in various formats – controlled environments or real-world scenarios – depending on your organisation's risk appetite and business context. It's crucial to ascertain whether your teams are genuinely ready for a cyber crisis. Too often, we hear reassurances like, "Yes, the backups should work." The word "should" is a red flag; certainty is the only acceptable standard when dealing with digital security.

Mature organisations recognise this and commit resources to identify and even test known vulnerabilities within their systems. This proactive approach allows them to gain first-hand insights into potential weaknesses. For example, after suffering a data breach, a leading tech company in the UK established a dedicated internal team to conduct regular penetration testing, actively seeking out and addressing vulnerabilities before attackers could exploit them.

To effectively train IT teams, start by analysing human and technical vulnerabilities by carrying out a thorough audit. This assessment sets the stage for intrusion teams to exploit these vulnerabilities during simulated attacks. Instructors will closely observe team reactions, latency times and operational performance, whether the test is purely theoretical or reflects actual conditions. This method provides concrete data on recovery times, restart rates and the speed of threat propagation.

Conducting this kind of training is no small feat, and it's common to encounter resistance from team members. Phrases like "I don't have time" or "We're already understaffed" may arise, highlighting the inherent challenges in making training a priority. However, without well-prepared IT teams ready to execute their roles in a cyber crisis response, the organisation is likely to face greater difficulties in the event of an incident, putting even more pressure on the CIO.

---

### Key Takeaways

In this chapter, we explored how crucial it is to maintain high performance within your IT and cyber security teams, particularly considering the increasing internal and external threats facing organisations today. As we prepare to move on to the next chapter, which focuses on the strategies for preparing these teams as the first line of defence, let's recap the key takeaways:

▶

▨ **Intelligence and Expertise Are Paramount**

A robust foundation of cyber intelligence, combined with skilled partnerships, is essential to stay ahead in the ever-evolving landscape of cyber defence. The integration of technical support, particularly through automation bolstered by artificial intelligence, plays a crucial role in enhancing the capabilities of your security teams. For example, organisations that utilise AI-driven threat detection can identify anomalies in real time, allowing for quicker responses to potential breaches.

▨ **Collaboration Is Key**

The IT teams within your organisation must be prepared to work seamlessly with their cyber security counterparts during a cyber attack. This collaboration is not just beneficial; it's essential for a coordinated response. Consider the 2020 Garmin ransomware attack, where swift communication between IT and cyber security teams was instrumental in managing the fallout and restoring operations efficiently.

▨ **Training Needs to Be Rigorous and Realistic**

Training for both IT and cyber security teams should be regular, realistic, and uncompromising. Cyber attackers are relentless, so your preparation should be too. Simulated-attack scenarios, which mimic real-world conditions, are vital for building resilience and ensuring that teams can respond effectively under pressure. Companies that conduct these exercises regularly, such as FireEye, have reported improved incident response times and a heightened state of readiness across their teams.

As we wrap up this chapter, it's clear that a high-performing IT and cyber security team is essential in safeguarding your organisation against a rapidly evolving array of threats. Building on this foundation, it's time to consider how these insights translate into actionable steps. Strengthening your defences doesn't end with theory, you need to put the principles into practice.

Now, let's look at some critical questions and the steps you can take to evaluate and enhance your cyber resilience. Use them as a guide as you fortify your teams and prepare for the realities of the cyber landscape.

**Call to Action**

Take a moment to reflect on your organisation's current approach to cyber security. Are your IT and cyber security teams equipped with the tools and training they need to respond effectively to emerging threats? Consider the following questions to guide your evaluation.

**Assess Your Readiness**

- How often do your teams engage in realistic training scenarios that reflect the current threat landscape?
- Are your incident response plans regularly updated and tested against real-world scenarios?

**Evaluate Your Collaboration**

- Are your IT and security teams aligned in their goals and strategies?
- What steps can you take to improve communication and collaboration between these critical functions?

**Prioritise Cyber Threat Intelligence**

- Is your organisation leveraging Cyber Threat Intelligence (CTI) effectively?
- How can you ensure that actionable insights from CTI inform your daily operations and long-term planning?

**Embrace Intelligent Automation**

- Have you considered implementing automation tools to enhance your team's capabilities?
- What specific tasks or processes could benefit from automation to free up your team's time for strategic initiatives?

**Commit to Continuous Learning**

- Are you providing ongoing training opportunities that challenge your teams to grow and adapt?
- How can you foster a culture of continuous improvement within your organisation?

By engaging with these questions, you can identify the immediate steps you should take to enhance your organisation's cyber resilience. Don't wait for a crisis to highlight vulnerabilities; take proactive measures today to fortify your defences. Remember, cyber security is not a destination but a continuous journey that requires dedication, collaboration and the willingness to adapt.

# 4

# PREPARING EMPLOYEES

> "Security is not a product, but a process."
>
> – Bruce Schneier

In the previous chapter, we detailed the importance of keeping IT professionals up to date with cyber-defence strategies. However, even the most skilled cyber protectors cannot single-handedly mitigate organisational risk if they cannot rely on the rest of the workforce being vigilant and prepared.

IT professionals act as the technical guardians of any organisation's digital assets, but employees in all departments play an equally critical role as defenders. This chapter explores how to prepare them effectively to serve as vigilant first responders to cyber threats. From creating a culture of awareness to implementing engaging and targeted training programmes, we will examine the strategies necessary to build a resilient front line, where each employee understands their role in safeguarding against cyber threats.

---

**Case Study: Mia Ash**

The case of Mia Ash, which dates to early 2017, underscores the importance of training employees to be vigilant and sceptical of unsolicited communication. Widely documented on platforms like Security Week,[1] this incident involved a group of attackers suspected to be associated with Iran, targeting companies across North Africa and the Middle East. The attackers combined advanced technological methods with social engineering to manipulate individuals, focusing particularly on male employees.

Their approach was simple yet effective: using a fabricated online persona named Mia Ash, they reached out to individuals under the guise of a young woman in need. Many employees, acting out of kindness or curiosity, engaged with her and opened

---

[1] Arghire, I., "Iran-Linked hackers use 'Mia Ash' honey trap to compromise targets", www.securityweek.com, 1 August 2017.

attachments to what was presented as her "photo book" from their business email accounts. This lack of scepticism allowed the attackers to infiltrate the companies' information systems.

This case highlights the critical role of training in building a culture of healthy suspicion towards unexpected or unsolicited communications. It's not about fostering paranoia but encouraging employees to practise cautious scepticism – an essential skill in the digital landscape.

500+
connections

**Mia Ash**
Photographer at Mia's Photography
London, Greater London, United Kingdom • **Contact info**

| | |
|---|---|
| Current | Mia's Photography |
| Previous | Loft Studios, Clapham Studios |
| Education | Goldsmiths, University of London |

bittersweetvenom24     **Follow**

**147 views**     23w

**Bittersweetvenom24** Jbam!
#oscarthefatcat #oscarthecat #sleepingcat #aww #boomerang

**Log in** to like or comment     ● ● ●

**LESSONS LEARNED**

This real-world example emphasises the value of security-awareness training that includes social engineering scenarios, equipping employees with the skills to question unusual requests and validate identities before engaging. Companies like KnowBe4 and Immersive Labs have developed adaptive training modules that help employees pick up on red flags in similar situations, ultimately reducing the risk of social engineering attacks.

# The First Line of Defence: Your Workforce

Employees in all departments are the first line of defence against cyber threats. As Benjamin Franklin wisely said, "An ounce of prevention is worth a pound of cure." This principle underpins the importance of equipping staff with the awareness and skills needed to spot and prevent potential threats. IDC's 2022 Future Enterprise Resiliency and Spending Survey[2] reinforces this, ranking security training as a top priority, with organisations investing in platforms like IBM, CybeReady, Immersive Labs, and KnowBe4.

### ▦ Training as a Priority

While cyber security teams focus on the technical side of protection, employees across the organisation play an essential role. Security training programmes are evolving to address this responsibility more effectively, using a mix of phishing simulations, social engineering scenarios, and advanced, AI-driven training modules that adapt to individual learning styles. These tailored programmes help employees retain knowledge and respond quickly to real threats, reinforcing their role as proactive defenders. Marcelo Nicácio, IT Executive Director at Tredegar, underscores this need: "Even with all the technology on our side, preparing employees to be the first line of defence when it comes to cyber security is essential. Ongoing preparation regarding threats and best practices helps to reduce internal and external risks." This training covers not just theory but hands-on exercises and a shared sense of responsibility, transforming employees into active contributors to the organisation's defences.

### ▦ Human Vulnerability in Cyber Defence

Despite sophisticated technical defences, human error remains one of the most exploited entry points for cyber attackers. The weakest link often lies "somewhere between the screen and the chair". For many employees, managing digital risks may not seem like a priority. Phrases like "I'm already pushed for time; cyber security just isn't part of my job" reflect a common mindset that attackers are quick to exploit.

Yann Bonnet, Deputy CEO of Campus Cyber, reminds us that "The issues of data security and the protection of IT services must be everyone's

---

[2] "Future enterprise resiliency and spending survey", www.idc.com, March 2022.

business. Each person needs to understand that their actions, if they lead to a cyber attack, can jeopardise the organisation and disrupt an entire eco-system." Just as seatbelts are second nature for drivers, employees must get into simple digital habits, such as verifying emails, updating passwords, and maintaining software. These basic precautions are powerful defences that can make all the difference.

### ■ Management's Role in Awareness

Creating a culture of vigilance starts with visible, proactive leadership. To encourage employees to go beyond simple compliance, managers need to model best practices and address cyber security openly. This goes beyond the annual training video and requires day-to-day discussions and a clear commitment to digital safety.

When employees consult their line managers about a potential security issue or demonstrate risky behaviour, managers must respond swiftly. By weaving cyber security into mission statements, codes of conduct, and everyday interactions, companies can show that digital safety is a shared priority. Some companies use memorable campaigns to reinforce these habits, with slogans like "Treat your passwords like your underwear – change them regularly, never share them, and keep them off your desk." A colourful approach to cyber hygiene!

# Oops, I Clicked It Again

The weakest link in cyber defence is often human behaviour. Many employees underestimate their role in digital security, an oversight that attackers eagerly exploit. For most, cyber security feels like an extra task on top of their daily work. However, this lack of concern about digital risk is exactly what cyber threat actors count on.

As Yann Bonnet, Deputy CEO of Campus Cyber, puts it, "The issues of data security and the protection of IT services must be everyone's business." He emphasises that individual actions can have a significant impact, poten-tially jeopardising not just the organisation but also affecting local networks and communities. The ripple effects of a cyber attack can last for months, underscoring the need for everyone to adopt safe digital habits, much like the instinctive habit of buckling a seatbelt. Simple actions – such as verify-ing email sources, scheduling regular backups, and updating software – can make a major difference to the overall security of a company.

> **Case Study: Twitter Hack of 2020**
> In 2020, a significant security breach at Twitter demonstrated the impact of social engineering on an organisation. Attackers successfully targeted Twitter employees, gaining access to internal tools through social engineering tactics. They persuaded employees to share login credentials, which allowed the attackers to access high-profile accounts and promote a cryptocurrency scam. This hack cost Twitter millions in financial and reputational damage and highlighted the critical importance of employee vigilance in mitigating cyber threats.
>
> **LESSONS LEARNED**
> Proper cyber-hygiene training and reinforcing policies that prohibit sharing credentials are essential. Regular simulations of social engineering attempts could help reduce the likelihood of employees falling prey to these tactics.

# Effective Preparation in the Real World

Improving the cyber security behaviour of employees whose primary responsibilities do not involve IT security is no easy task. It requires time, dedication, and an understanding of how management behaviours influence staff actions. Despite efforts to mitigate risks, statistics show that an average of 10% of employees will always fall victim to social engineering attacks, including phishing.

To effectively train employees, organisations can draw from principles of Murphy's Law, which suggests that anything that can go wrong will go wrong, especially during high-stress periods. Consider the following points:

**Timing Is Key**

Attackers often choose moments of heightened stress to launch their campaigns, such as during busy sales periods or after a major corporate announcement. Preparing your teams to recognise threats during these critical times enhances the effectiveness of training. For instance, a sales team may receive a phishing email disguised as a request from a client just before their year-end closing.

**Conducting Simulated Attacks**

Implementing phishing tests and email-compromise drills during peak business times can create a more impactful learning experience. Employees who receive plausible requests during high-pressure periods are more likely to engage with them, providing valuable teaching opportunities.

■ **Regular and Progressive Training**

Annual certifications for all employees can provide a foundation for under-standing digital risk, but it's essential to provide ongoing training tailored to the specific challenges each division faces. Use progressive difficulty levels in training exercises to ensure employees are challenged appro-priately. Start with easily recognisable phishing attempts and gradually increase complexity based on their performance.

■ **Leadership Involvement**

Managers must actively participate in the training process. By addressing cyber security issues and demonstrating a commitment to safe practices, they can motivate employees to take these matters seriously.

■ **Creating a Culture of Awareness**

Establishing a culture where cyber security is everyone's responsibility requires consistent messaging and visible leadership. Employees should feel empowered to report suspicious activity and ask questions about security protocols without fear of reprimand. As Alexandre Gazzola, Regional Director at Orange Cyberdefense, aptly states, "All cyber secu-rity experts today agree that people have an important role to play in cyber defence. Our investigations confirm that most cyber-attack chains include one or more human links." This acknowledgment of human vulnerability is crucial in shaping training and awareness programmes.

## Running up That Hill

Let's face it, with the best will in the world, between 8% and 12% of diehards will always click on the wrong link or open a dodgy attachment. When you think about the number of people – and the number of gateways into the information system – it's a steep and challenging hill to climb. Keep this front of mind; an effective cyber-defence system is made up of layers combining tools, skills, and culture.

### Understanding Risks: A Matter of Staying Sharp

Think of cyber security awareness like walking along a crowded city street with your wallet in hand. Would you wave it around openly, inviting anyone to grab it? Probably not. The same goes for digital security – it's about knowing when to keep a close grip on sensitive information and being cautious about what you share and where.

Just as you wouldn't leave your front door unlocked, don't "leave the door open" online by overlooking suspicious messages or unfamiliar links. According to a report by Verizon,[3] over 90% of successful cyber attacks begin with a phishing email. A few extra seconds of vigilance can prevent a lot of hassle down the line.

### Locking the Digital Door

Framing digital habits as personal safety steps can foster a mindset where daily actions add up to a strong defence. According to a KnowBe4 study,[4] 70% of employees who received regular security awareness training reported feeling more confident about recognising threats. Making cyber security second nature – like checking the door before you leave – can significantly reduce incidents. A little caution now saves a lot of trouble later.

---

**Case Study: Deepfake Attack on a UK Energy Company**

In 2019, an energy company in the UK fell victim to a sophisticated cyber attack involving deepfake technology. Attackers used AI-generated audio to impersonate the CEO's voice, instructing an executive to transfer €220,000 to a fraudulent account. The accuracy of the impersonation was so convincing that the executive complied without question. This attack underscores the importance of verification protocols, especially when instructions seem urgent.

**LESSONS LEARNED**

Encourage employees to verify unusual requests through multiple channels, particularly when dealing with financial transactions. Technology and training that emphasises multiple authentication steps can help mitigate such risks.

---

# Measuring the Impact of Training

How can you determine whether a training session is effective? Measuring the impact of cyber security training is essential to justify continued investment in both human and financial resources. Effective training can be evaluated by focusing on specific metrics, particularly around reducing the number of basic incidents reaching cyber security teams and supporting risk-mitigation efforts.

---

[3]  Sheridan, K., "85% of data breaches involve human interaction: Verizon DBIR", www.darkreading.com, 13 May 2021.
[4]  "2021 Cybersecurity awareness training benchmarks", KnowBe4.

# Key Metrics to Assess Employee Preparedness

To gauge the effectiveness of cyber security training programmes, organisations commonly track several metrics:

■ **Incident Reduction**

A measurable decrease in incidents reaching Security Operations Centre (SOC) or Computer Emergency Response Team (CERT) levels. This indicates that employees are handling simple issues independently, reducing the load on specialised teams.

■ **Detection and Response Time**

Implementing regular simulation training and phishing drills has been shown to significantly improve detection and response times. For instance, organisations that use platforms like Immersive Labs have achieved a 30% reduction in incident response time, enabling swift responses that minimise damage and keep business operations running.[5]

■ **Success Rates in Phishing Simulations**

Tracking click-through rates on phishing simulations provides insight into employees' risk awareness. A typical goal is to reduce this rate significantly over time. KnowBe4 found that companies with continuous training see up to a 75% drop in phishing susceptibility.[6]

■ **Reporting Rate**

The percentage of employees reporting suspicious activity post-training, as this signals an increase in vigilance and proactive security behaviour.

**Example Table**   Metrics for Gauging Training Effectiveness

| Metric | Purpose | Target Improvement |
| --- | --- | --- |
| Incident Reduction | Measures training effectiveness in preventing simple issues from escalating | Decrease incidents by 20–30% over a year |
| Detection and Response Time | Reduces the response time for identifying and handling incidents | 25–30% faster response post-training |
| Success Rates in Phishing Simulations | Tracks phishing awareness and response | Reduce click-through rates by 50–75% |
| Reporting Rate | Increases proactive behaviour in threat detection | 60–80% increase in reporting |

---

[5] "Immersive Labs global study finds improved response time to threats, yet resilience efforts still fall short", https://www.immersivelabs.com, 2 August 2023.
[6] "2021 Cybersecurity awareness training benchmarks", KnowBe4.

# Simulated vs. Real Attacks: How They Inform Training Effectiveness

Understanding the difference between simulated attacks (e.g., phishing tests) and real incidents is key to assessing preparedness.

■ **Simulated Attacks**

These exercises, conducted at regular intervals, allow teams to measure initial awareness and adaptability. Positive responses, such as declining click-through rates on phishing simulations, indicate growing security awareness. Organisations new to digital-risk mitigation may start with around 75% of employees adopting good habits, whereas mature companies often achieve levels closer to 95% in training assessments.

■ **Real Incident Metrics**

Metrics from real cases handled by SOC (Security Operations Centre) or CERT (Computer Emergency Response) teams provide a direct measure of employee readiness in live scenarios. If basic incidents, like suspicious emails or low-complexity phishing attacks, decline in number over time, it suggests that training is taking effect. In IBM's "2021 Cost of a Data Breach" study, organisations with comprehensive training programmes reported an average cost reduction of $1.76 million in data breaches, attributed partly to employees' enhanced ability to handle lower-level incidents independently.

Success in employee preparedness is gauged through this combination of reduced incident volume at higher security levels and quicker, more accurate responses from employees.

---

**Key Takeaways**

As we learned from this chapter, employees – regardless of their role – are the gatekeepers of effective cyber defence. Before we move on to the next chapter, which will discuss how to prepare general management, here's a summary of the main points we have covered:

■ **Human Factors**

Cyber defence begins with individuals. Awareness is crucial for everyone. Even a small lapse in vigilance can lead to significant risk exposure, so every employee plays a role in safeguarding the organisation.

■ **Ongoing Vigilance**

Cyber vigilance is a practice, not a one-off event. Just as any skill requires consistent exercise, maintaining security awareness involves regular reinforcement, repetition and updating to align with evolving threats.

■ **Targeted Interventions**

Focus training where it's needed most. Regular, focused training ensures resources are allocated effectively, addressing areas of highest vulnerability and promoting efficient use of time and resources.

By embedding these principles into daily routines, building muscle memory, and fostering a culture of shared responsibility, organisations can empower their employees to act as a unified, proactive line of defence against cyber threats.

# 5

# PREPARING SENIOR MANAGEMENT

> "In preparing for battle, I have always found that plans are useless, but planning is indispensable."
>
> – Dwight D. Eisenhower

Cyber threats are as unpredictable as any battlefield. Effective preparation builds the skills and reflexes required to lead under pressure, while planning empowers leaders to make decisions that sustain resilience. As strategic decision-makers, senior executives and board members are responsible for guiding organisations through today's evolving digital threats. This chapter explores how targeted crisis training equips them with the skills, insights and confidence to act decisively when it counts.

## Crisis Training for Leaders – Risk Awareness

The advantages of crisis training go far beyond immediate response readiness. Leaders develop a practical understanding of digital threats through exposure to realistic scenarios. This training allows them to anticipate cyber criminals' tactics, understand potential risks and better assess their organisation's vulnerabilities. By thinking like an attacker, executives gain insight into areas that may be exploited, enhancing their ability to safeguard critical assets.

### ■ Team Cohesion

Crisis training promotes collaboration and unifies leadership in all departments. By establishing a shared language and set of priorities around security, leaders foster a cohesive, company-wide commitment to cyber resilience. Cross-departmental teamwork is strengthened, creating a united front that can execute a swift and efficient crisis response.

■ **Regulatory and Law Enforcement Partnerships**

Training also builds connections with regulatory bodies and law enforcement. These partnerships are vital for crisis response and post-incident management, especially when compliance and regulatory reporting are required. For example, maintaining close ties with relevant authorities can expedite support and guidance during a crisis, helping leaders manage both the immediate response and any subsequent inquiries.

■ **Building Trust by Staying Prepared**

Investing in cyber crisis training demonstrates a proactive commitment by senior leadership to security and effective risk management. This priority fosters trust with stakeholders, clients, and the broader public. By taking visible, preventive steps to protect data and ensure business continuity, leaders strengthen their relationships with customers, partners, and investors.

In industries like finance and healthcare – where data protection is a high priority – these efforts can significantly enhance public perception and confidence. Clients and stakeholders see that management is dedicated to security as a long-term investment, creating a secure environment in which both the organisation and its customers can thrive.

Consider the perspective of Mathieu, Group CISO at a major investment bank in France: "Companies face several challenges – regulatory pressures, the need to adopt new technology, and maintaining legacy systems. Boards and executives need a clear understanding of these issues and how they impact security. Crisis exercises help board members understand how the organisation's systems work, where the risks are, and what it takes to manage them. This understanding is invaluable."

# Defining Cyber Resilience

Cyber resilience refers to an organisation's ability to not only withstand but also adapt and recover from cyber incidents with minimal disruption. This resilience is no longer just a defensive measure; it's a strategic necessity. A successful cyber attack can have far-reaching effects that impact both operations and reputation. Resilience helps organisations reduce these impacts and get back to business quickly.

## The Core Role of Senior Management in Cyber Resilience

Senior management plays a critical role in shaping a resilience posture by integrating cyber security with their company's strategic goals. This integration empowers leaders to make security a priority that supports business continuity and sustainable growth.

- Resilience as a Business Imperative: Cyber resilience enables companies to protect operations, maintain trust, and support long-term growth.
- Rapid Recovery and Continuity: By embedding resilience into the organisation, leaders ensure that key systems and processes can withstand cyber incidents and recover swiftly.
- Trust and Credibility: A strong security commitment reassures stakeholders, creating a durable competitive advantage in industries where security is a top concern.

When a business invests in cyber crisis training for both management and employees, it enhances its long-term stability. Resilience is beneficial to everything from operational continuity to preserving customer trust. By prioritising cyber security at every level, a company demonstrates its commitment to sustainable success and shows stakeholders that it is well prepared to face future challenges.

## Where Security and Resilience Meet: A Modern Love Story

Security and resilience must be viewed as inseparable; together, they lay the foundation for sustainable growth. To build a robust framework, organisations should:

- **Proactively Commit to Robust Cyber Security**

  By investing in strong cyber security measures, aligned with business objectives and risk tolerance, a company can ensure that security is not only a reactive measure but a proactive stance. By aligning security practices with strategic goals, leaders ensure that the organisation is well positioned to respond to evolving threats and maintain operational stability.

- **Implement and Continuously Refine Security Protocols**

  Resilient security protocols need regular testing and updates to stay agile and adaptable to changing threats. A commitment to rigorous, continuous refinement minimises vulnerabilities and builds a responsive, flexible security posture that can withstand disruptions without compromising essential functions.

The concepts of security and resilience are two sides of the same coin. An insecure environment is unsustainable. In today's era of rapid digital innovation, organisations that prioritise resilience at the core of their operations build a foundation for long-term success. This commitment to resilience inspires stakeholder confidence, cultivates lasting trust, and enables them to navigate cyber incidents with confidence and agility.

The future belongs to organisations that fortify the link between security and resilience. By integrating strong security practices into the company's fabric, it gains the durability to achieve sustainable growth and show an unwavering determination to withstand the relentless tide of cyber threats. Security isn't just an add-on; it's a fundamental pillar of organisational strength that helps businesses establish a legacy in a challenging digital landscape.

## Embracing the Culture

As organisations embrace innovative technologies, they must also confront the cyber risks that inevitably follow. Increased connectivity – from digital tools to the Internet of Things (IoT) and integrated operational technology (OT) – has created an environment in which technological progress and cyber risk go hand in hand. Companies cannot simply unplug from the digital world and, as a consequence, cyber resilience has become indispensable.

Nicolas Meyerhoffer, Vice President of IBM France, captures this idea: "Technological progress and the digitisation of businesses are consubstantial with cyber risks. The incredible enthusiasm for generative AI is just further proof of this: its potential is extraordinary, and yet it brings a new risk that needs to be controlled. The more significant the technological advance, the greater the risk. In this context, the role of managers is to give companies the impetus they need to derive maximum benefit from innovations while supplying the means to keep them under control, which includes useful cyber security control capabilities."

To successfully innovate while maintaining resilience, companies can "embed cyber security into their approach" through methods like automated threat-detection tools, which proactively monitor for suspicious activity, and regular vulnerability assessments, identifying and mitigating potential risks before attackers can exploit them. With these practical steps, organisations not only advance technologically but also maintain robust defences to withstand emerging threats.

## Getting It Right

Cyber crisis training is in high demand, with some companies booking sessions months in advance. For senior executives, digital-risk mitigation has evolved from a concern to a strategic priority. Crisis training held every six to twelve months helps keep leaders' crisis-response skills sharp, readying them to act decisively and effectively.

A structured approach like the RPA Model (Repetition, Precision, Anchoring) is essential for effective training:

■ **Repetition**

Repetition builds reflexes, enabling leaders to respond instinctively in high-stress scenarios. By repeatedly practising crisis responses, leaders internalise key actions and are more likely to act with confidence in a real crisis.

■ **Precision**

Precision involves creating tailored, scenario-specific exercises that reflect the organisation's unique environment. This ensures that leaders practise in situations that mirror their actual operational landscape, reinforcing the relevance of training and building confidence.

■ **Anchoring**

Anchoring solidifies crisis-response skills, establishing them as habits. This element of the RPA Model is crucial for creating long-term resilience, as each exercise builds on prior training, reinforcing learned skills and responses over time.

## The Secret Sauce

Effective cyber crisis training offers a structured process to evaluate a company's readiness for real-world threats. Here's how this process unfolds:

■ **Realistic Replica Scenarios**

Scenarios are crafted to replicate real-world crises, challenging leaders to respond as they would in an actual event. Complexity increases with each level, starting with basic exercises that include external stimuli like phone calls and press releases and progressing to advanced simulations involving larger portions of the company. For example, during a training session with a well-known manufacturer, a real-world incident involving the company's e-commerce site coincidentally mirrored the simulated scenario, reinforcing the realism of the exercise.

■ **Decision-Making and Problem Solving**

Participants are tasked with making quick decisions, prioritising actions, and collaborating with others in high-pressure, real-time situations. Experienced clients use their existing protocols, while less mature organisations receive guides on what steps to take. This approach strengthens critical decision-making skills and problem-solving abilities, which are essential for navigating a cyber crisis.

■ **Observation, Communication and Coordination**

Simulations highlight the importance of clear communication and effective coordination among leaders. Each session begins with an elevated stress level, assigning one person the role of coordinating responses across business lines. This setup demonstrates the necessity of leadership and communication in a crisis, preparing executives to guide their teams and make efficient, cohesive decisions.

■ **Emotional Impact**

Crisis simulations evoke real emotional responses, especially for executives less accustomed to managing cyber risks. This experience leaves a lasting impression, driving home the urgency of proactive risk management. Trainers carefully balance the intensity of the simulation, ensuring participants feel the pressure without becoming overwhelmed. This level of immersion helps leaders internalise lessons, fostering reflexive responses for real-world crises.

## Turning up the Heat

As a company matures in its approach to security, crisis-training exercises should evolve to match, becoming more challenging and comprehensive. Here's how training scenarios can progressively build leaders' crisis management skills at different levels:

■ **Foundational Level: Establishing Basics**

In an initial simulation, leaders might respond to a mock phishing attack targeting internal systems. This entry-level exercise helps participants familiarise themselves with established protocols, emphasising quick identification and response in a low-pressure setting. This level builds foundational skills and sets the stage for more complex scenarios.

■ **Intermediate Level: Building Coordination and Communication**

At this stage, a simulated ransomware attack could be introduced, requiring leaders to coordinate responses across departments and engage with external partners. Leaders are tested on cross-functional collaboration, quick prioritisation and communication management as they work together to contain the threat and maintain continuity.

### ▓ Advanced Level: Comprehensive Crisis Response

Leaders at the most mature level could face a multifaceted simulation where both digital and physical operations are compromised. They must respond to disrupted production lines, handle media inquiries and maintain communication with regulatory authorities. This full-spectrum scenario provides a realistic test of leadership under intense pressure, reinforcing complex decision-making and team coordination.

By advancing the complexity of crisis simulations as the organisation's security posture matures, leaders gain a deeper understanding of crisis management and develop stronger, more adaptive responses. This gradual approach helps identify and address vulnerabilities, ultimately strengthening the organisation's resilience against real-world cyber threats.

---

**Case Study: The TV5Monde Cyber Attack**

In April 2015, TV5Monde, a prominent French television network, experienced a significant cyber attack that had widespread repercussions. Hackers claiming allegiance to the Islamic State group, under the alias "CyberCaliphate", breached TV5Monde's systems, halting its broadcasting for several hours and defacing its social media accounts with propaganda. The disruption affected operations and caused substantial reputational damage.

Upon investigation, cyber security experts attributed the attack to the Russian hacking group APT28, also known as "Fancy Bear," suggesting that this was not a lone act of digital vandalism but a sophisticated, state-sponsored effort. This incident underscored the vulnerabilities within the media industry and highlighted the profound impact that targeted cyber attacks can have on critical infrastructure.

**LESSONS LEARNED**

- ▓ The Role of Preparedness: The attack on TV5Monde served as a wake-up call for many industries and illustrated the importance of proactive crisis management. Effective training builds reflexes and equips organisations to respond swiftly and decisively.
- ▓ Reputational and Operational Risks: With its operations disrupted and public image impacted, TV5Monde exemplified how cyber incidents extend beyond IT concerns and can severely damage organisational credibility and customer trust.

Jacques, CISO of a prominent media organisation in France, reflected on the attack: "The best crisis response is preparation well before the crisis hits. This training builds reflexes, ensuring that everyone – from operational staff to top executives – is prepared." This case reinforces that a robust, well-practised response strategy is invaluable, especially in high-stakes industries like media.

---

## The Devil Is in the Detail

Effective crisis simulations rely on deep immersion and attention to detail, which are essential for keeping participants engaged and fostering realistic experiences. Even a minor inconsistency, such as an unrealistic news report or an out-of-place communication channel, can break the immersion and cause senior management to disengage.

## Creating High-Pressure, Realistic Environments

Deeply immersive simulations create the intense, high-pressure environment that mirrors an actual cyber crisis. To achieve this level of realism, elements often include:

■ Simulated News Reports and Social Media Activity: News reports and social media responses are crafted to appear as they would in a real-world crisis, reflecting the likely public reaction and media scrutiny. This helps executives anticipate how their decisions could be perceived externally (see Pocket Guides 4 & 5).

■ Phone Calls from "Customers" and "Authorities": Simulated calls from key stakeholders like customers, partners, and regulatory authorities add an element of real-time decision-making, pressing leaders to respond quickly and effectively.

■ Press Articles and Fake Websites: Customised press articles and mock websites heighten the stakes, giving leaders insight into how quickly information spreads – and how the company's reputation may be affected.

■ Ransom Demands and Deepfake Media: Incorporating elements like ransom demands, deepfake videos, and audio (see Pocket Guide 6) intensifies the scenario, requiring executives to make swift, sound decisions under pressure, just as they would in a real attack.

When facilitators incorporate these immersive elements into the training, participants can fully engage in the exercise, experiencing the urgency and gravity of a cyber crisis. This leads to more practical, insightful lessons and ultimately improves their real-world crisis response.

# The Bespoke Fit

Effective crisis training is built on a custom approach that mirrors an organisation's specific needs and challenges. This bespoke approach involves several key steps:

■ Partnering with IT and Security Teams: Close collaboration with IT and security teams enables simulations to accurately reflect the company's unique vulnerabilities and system specifics, aligning each scenario with actual risk areas.

- Engaging Senior Leadership: Involving senior leaders helps tailor simulations to the organisation's strategic vision and decision-making style, creating realistic scenarios that mirror true leadership dynamics and communication patterns.

- Aligning to Key Objectives: Each simulation is designed to align with the organisation's current business priorities and regulatory landscape, making the training relevant and directly applicable to day-to-day operations.

This tailored approach fosters engagement and ensures that every department understands its role in a coordinated response. By aligning each team's role within the simulation, companies will strengthen their resilience and build a culture of readiness, empowering all employees to respond effectively, even under pressure.

## Meet the Team

Building strong relationships with external partners strengthens a company's crisis-response capability, ensuring that the right expertise is accessible when it matters most. These partners, which may include insurers, ransom negotiators, regulatory authorities or technical responders, each play a critical role in a well-coordinated crisis response. By involving them in advance, organisations lay the groundwork for an efficient, unified response that minimises damage.

Key Partners and Their Roles:

- Insurers: Cyber insurers provide more than financial coverage; they offer valuable insights on risk management. In a ransomware situation, for example, they may advise on best practices for mitigation, guide on whether ransom payment is covered under the policy, and provide connections to additional support, such as ransom negotiators.

- Ransom Negotiators: Experts in managing high-stakes situations, ransom negotiators have the experience needed to engage attackers and de-escalate demands. One European tech company, for instance, includes its ransom negotiator in crisis training to ensure alignment with management's expectations and to build rapport with key stakeholders before a real crisis occurs.

- Regulatory Authorities: Engaging with regulatory bodies in advance establishes clear channels for compliance, reporting, and external support during a cyber crisis. Authorities can provide timely updates on emerging threats and advise on managing legal exposure. Additionally, their involvement in simulations reinforces regulatory compliance and trust with the organisation.

■ Technical Responders: External incident responders bring specialised skills and tools to contain and resolve the technical aspects of an attack. Their involvement in simulations builds familiarity with the company's systems, allowing them to respond quickly when they are needed.

> *"Good management of subcontractors is essential."*
>
> – Frank Van Caenegem

Frank Van Caenegem, Cybersecurity Vice-President and CISO EMEA at Schneider Electric, emphasises, "Good management of subcontractors is essential. It's not only about their reliability but also understanding the exposure we have due to our dependence on them." Van Caenegem's point highlights a vital takeaway: preemptive planning with key stakeholders is essential to establish trust, refine procedures, and synchronise expectations. Conducting joint simulation exercises fosters collaboration, enabling everyone involved to respond smoothly during a real crisis.

## The Perennial Pest

Ransomware attacks (see Pocket Guide 1), which have become increasingly common, require organisations to approach ransom demands strategically and with caution. Each scenario presents unique challenges, often forcing leaders to weigh the risks and rewards of potential payment options.

### Key Considerations for Ransom Negotiation

■ Legal and Regulatory Implications: Consulting legal teams before any ransom payment is crucial. In some jurisdictions, paying a ransom may violate laws, particularly if the attacker is linked to terrorism. Working with law enforcement can also help confirm whether the attackers are on restricted lists.

■ Assessing Impact: Each decision should be made case by case, assessing factors like the severity of operational disruptions, the potential financial loss, and, most critically, the safety of human lives. Companies should have backup plans in place, including data recovery and business-continuity procedures, for cases where payment isn't an option.

■ Reputation and Long-Term Effects: Companies that opt to pay ransom may risk signalling vulnerability to other attackers. For this reason, ransom discussions should only be initiated by the attacker to avoid projecting desperation. Consult with senior management, legal advisors, and crisis teams to determine the most appropriate response.

## Practical Steps

■ Involve Legal Teams: Always consult legal teams prior to making any payment decisions to ensure compliance with local and international laws.

■ Engage with Ransom Negotiators: If negotiation is considered, delegating this task to experts skilled in de-escalation and communication with attackers is advisable. Professional ransom negotiators, often from law enforcement backgrounds, understand the complexities of managing such sensitive conversations.

■ Have Backup Plans: Prepare for alternative solutions by strengthening data recovery protocols and backup systems to minimise reliance on ransom payments.

---

**Case Study: WannaCry**

The WannaCry ransomware attack in 2017 provided a stark reminder of the global risks associated with unpatched vulnerabilities. Exploiting the EternalBlue vulnerability in Microsoft Windows – originally developed by the US National Security Agency (NSA) – the attackers rapidly spread ransomware in over 150 countries, impacting diverse sectors including healthcare, logistics, and public services. The UK's National Health Service (NHS) was particularly affected, experiencing widespread disruption to patient care and essential services.

**LESSONS LEARNED**

■ Global Collaboration Is Essential: The WannaCry response highlighted the importance of international cooperation. Cyber security firms, law enforcement agencies, and governments worldwide worked together to share intelligence, monitor Bitcoin transactions, and develop decryption tools. This coordinated approach underscored the value of public–private partnerships in managing large-scale cyber incidents.

■ Proactive Security Measures Prevent Catastrophes: WannaCry demonstrated the critical role of timely software updates. Organisations that had applied the necessary patches were protected, showing the importance of vulnerability management and regular security updates.

■ Preparedness Accelerates Recovery: Rapid response strategies, including backup systems and incident response protocols, enabled affected organisations to minimise disruption. Many cyber security firms quickly provided decryption tools so that data could be recovered without paying ransoms. This reinforced the need for organisations to have well-prepared incident response plans to reduce recovery time during cyber crises.

The WannaCry attack illustrated the importance of global collaboration, proactive security measures, and the need for rapid responses in mitigating the damage caused by ransomware attacks.

## External vs. Internal Instructors

In crisis-management training, the choice between external third-party facilitators and internal instructors can significantly impact the effectiveness and depth of the simulation. Each approach offers distinct advantages; understanding their roles can help companies create a more resilient response framework.

### Advantages of External Third-Party Facilitators

■ **Impartiality and Unbiased Feedback**

External facilitators bring an objective, unbiased perspective that is invaluable during crisis training. Free from internal dynamics or organisational culture, they offer impartial feedback, allowing teams to gain an honest assessment of their strengths and areas for improvement. This objectivity helps executives recognise critical gaps without concerns of internal bias or influence, enhancing the accuracy and effectiveness of training.

■ **Broad Industry Insights and Best Practices**

With experience across sectors and industries, external facilitators introduce a wide array of best practices and innovative strategies that internal teams may not be exposed to. This cross-industry perspective enriches the training by providing fresh approaches, adapting lessons learned in different environments and sharing insights into evolving threats that might be less familiar to existing internal teams.

■ **Benchmarking and Industry Standards**

External instructors offer organisations the chance to benchmark their preparedness and crisis-response capabilities against industry norms. Through anonymous comparisons with similar organisations, they can highlight areas where the company excels or falls short to gain a better understanding of where they stand within their industry. This benchmarking helps leaders prioritise improvements relative to their peers.

- **Current Threat Awareness and Adaptability**

  External instructors are often deeply connected to the evolving cyber landscape, bringing current knowledge of threats, tactics, and emerging risks. This ensures that training remains relevant, equipping participants with insights into new and potential cyber risks, such as the latest ransomware variants or social engineering tactics. This adaptability helps organisations remain proactive as cyber threats shift.

- **Enhanced Credibility and Confidence**

  The presence of qualified, experienced third-party facilitators lends credibility to the training process. Their impartiality and proven expertise can inspire confidence among participants, ensuring that teams feel they are learning from the best in the field. This reassurance can drive deeper engagement and trust in the training, helping participants internalise the lessons.

As Sébastien, Executive Assistant to the French Managing Director of a major IT company, remarks: "The simulations run by external experts are invaluable. They demonstrate to board members that a company's response to a cyber attack goes beyond procedural steps. Preparing the organisation requires ongoing, independent insights, and resilience is a muscle that needs constant exercise."

## Advantages of Internal Instructors

- **Deep Knowledge of Organisational Systems and Culture**

  Internal instructors possess intimate knowledge of the company's systems, processes, and cultural dynamics, allowing them to tailor training exercises precisely to the internal environment. They understand the specific protocols, business models, and departmental interactions that external trainers may need time to learn. This familiarity enables them to tailor scenarios to the company's existing frameworks and nuances.

- **Cost Efficiency and Long-Term Integration**

  Using internal resources for crisis training can be more cost-effective than bringing in external facilitators, especially for ongoing programmes. Internal instructors can integrate crisis-management training into regular operational practices, making it part of an ongoing resilience-building culture.

This approach is particularly useful for companies with tight budgets or those looking to develop a long-term, sustainable training programme.

■ **Rapport with Teams**

As they are part of the company, internal instructors already know the participants. This familiarity can foster a comfortable learning environment where employees feel more open to engage, ask questions, and reflect honestly. These relationships can help create a collaborative atmosphere conducive to effective training, especially when team building is a focus.

■ **Alignment with Organisational Goals**

Internal trainers are often deeply aligned with the company's goals, values, and mission. They can ensure that simulations and training exercises are structured to support organisational priorities, whether that's a specific regulatory focus, unique industry standards, or core strategic objectives. This alignment helps ensure that their training is practical, applicable, and directly relevant to the organisation's long-term vision.

### Combining Both Approaches for Maximum Impact

A blended approach, using both external third-party facilitators and internal instructors, can provide a comprehensive training experience. External facilitators can bring impartiality, industry insights, and updated threat intelligence, while internal instructors add continuity, organisational alignment, and a cultural fit. Companies that balance both perspectives will benefit from a well-rounded, resilient training programme, and their leaders will be as prepared as possible to handle even the most complex crisis scenarios.

## Words in Action

Effective communication is essential for managing any crisis, especially in high-pressure cyber incidents. Clear and coordinated messaging is needed across all levels – internal, among leaders, and externally – to preserve trust and ensure rapid, organised responses.

### Internal Stakeholder Communications

In a crisis, internal communication is critical for maintaining stability and morale. Employees are essential assets who contribute skills, insights, and resilience. "Leader's Intent" is a communication strategy where leaders provide guidance on the end goals without micromanaging each step, empowering teams to act decisively in the moment. For instance, during the NotPetya attack, Maersk's leadership used Leader's Intent with the guidance, "Do what you think is right for the customer; don't wait for headquarters." This approach can be adapted across industries to show employees that leadership trusts their judgement while providing a clear mission.

**Practical Example:** In a financial services firm managing a client data breach, leaders used Leader's Intent by encouraging employees to focus on safeguarding client data and trust. This clarity helped teams prioritise tasks and act swiftly, reducing downtime and restoring confidence.

## Leader-to-Leader Communications

Crisis management among senior leaders requires cohesion and efficient decision-making. When leaders communicate openly and respectfully, it creates alignment and helps them manage the crisis with minimal friction. Structured roundtable discussions and clear role assignments can ensure that decision-making remains organised and that leaders remain unified.

**Actionable Strategies:**

- Roundtable Decision-Making: Establish a roundtable format for discussions where leaders share updates and make collective decisions to reduce miscommunication.
- Role Clarity: Define roles ahead of time so each leader understands their responsibilities, helping avoid overlap and ensure swift action.

**Practical Example:** In a cyber crisis exercise, a technology firm practised leader-to-leader communication by assigning specific roles for decision-making, risk assessment, and external messaging. This approach ensured that all leaders had a clear function, improving efficiency in real-time responses.

## External Stakeholder Communications

In high-profile crises, external messaging must be consistent with internal communications to maintain trust. Aligning messages for stakeholders – clients, regulators, investors, and the public – is essential to uphold the organisation's reputation and mitigate potential backlash.

To streamline this process, companies can prepare communication templates that outline key messages for different crisis scenarios. These templates can then be customised based on the specific details of the incident.

**Practical Example:** In the 2019 Capital One data breach, the company's transparency in sharing key details of the incident and the steps it was taking to safeguard data helped to reassure stakeholders, preserving customer confidence despite the breach.

**Key Takeaways**

Senior management has a vital role in navigating cyber crises and setting the tone for resilience. Here's a summary of the critical lessons:

■ **Leadership in Crisis**

Senior leaders must exemplify calm and decisive action under pressure, as "the body follows the mind". When leaders project confidence, it inspires trust and motivates the entire organisation to stay focused.

■ **Realistic, Tailored Training**

Crisis training should go beyond standard protocols. Immersive, customised simulations build instinctive responses tailored to the organisation's unique environment, preparing leaders to adapt dynamically to any challenge.

■ **Engaging Professional Instructors**

Bringing in expert, independent instructors enhances training with industry-wide insights, constructive feedback, and realistic scenarios. Skilled instructors balance intense pressure with support, helping leaders build critical skills in a controlled environment. As the saying goes, "It's unwise to provoke someone well-prepared for the challenge."

This holistic approach strengthens organisational resilience, enabling teams to face crises with agility and confidence, while fostering a culture of preparedness that sustains long-term security.

# 6

# CYBER CRISIS COMMUNICATION

> "The single biggest problem in communication is
> the illusion that it has taken place."
>
> – George Bernard Shaw

The purpose of this chapter is to focus on a specific aspect of cyber crisis preparedness: communication. But why dedicate an entire chapter to crisis communication in a book covering all the organisational pillars of cyber crisis preparedness? Based on our experience, effective crisis management is, above all, about communication: knowing your environment, mobilising your team and partners, guiding stakeholders, staying calm, managing the narrative, and adapting to the attacker's messaging. Communication, communication, and more communication.

Unlike standard communication, which is aimed at shaping public image and promoting products or services, crisis communication is rooted in managing perception. It aims to prevent a reputational crisis from worsening the crisis already underway. Recent events, such as the COVID-19 pandemic and the Colonial Pipeline cyber attack in 2021, highlight just how critical it is to communicate clearly in times of crisis – and why it's vital to prepare thoughtfully in advance. However, do all the principles of traditional crisis communication apply the same way in a cyber crisis? Yes and no. In this chapter, we'll examine the similarities and unique demands of cyber crisis communication, drawing on real-world cases and expert insights to underscore practical approaches.

To understand the impact of crisis communication, let's first look at an example from the COVID-19 pandemic. Governments, companies, and health institutions had to constantly adapt their messages to respond to a rapidly evolving situation, adjusting to new knowledge about the virus and changing directives for public health. The pandemic showed us what happens when there's a flood of contradictory information: trust is shaken, and confusion spreads. With official messages and differing opinions circulating in media and social networks, this crisis exposed the importance – and the complexity – of clear, coordinated messaging.

What about cyber crisis communication? Consider the case of Colonial Pipeline in May 2021, when a major cyber attack paralysed the US oil transport company's distribution network, leading to fuel shortages and rising public concern. Crisis communications were immediately set up to keep the public informed about the attack's impact and recovery efforts, with support from the highest levels of the US government. President Joe Biden reassured the nation, stating, "Federal agencies moved quickly to mitigate the impact on our fuel supply." This incident highlighted the need for transparency, empathy, and honesty, showing that companies must be ready to communicate proactively to reassure stakeholders.

Yet the Colonial Pipeline case also reveals some key distinctions in cyber crisis communication. Here, the attackers publicly apologised for the broader impact of their actions, even as they continued their attack – a rare development compared to traditional crises. Unlike a natural disaster, where harm is often impersonal, cyber crises involve intelligent, motivated, and organised attackers whose actions are intentional and sometimes public. This dynamic adds another layer of complexity to communication, requiring organisations to address concerns beyond immediate operational impact.

Whether dealing with physical, financial, or digital crises, it's essential to prepare communication strategies in advance. Crises force communicators into unfamiliar territory, demanding that they respond quickly, often amid intense public scrutiny. Unlike peacetime messaging, which is strategic and long-term with goals like brand growth and positive image building, crisis communication must be reactive and tactical, focusing on managing perceptions, reassuring stakeholders, and minimising damage.

In our experience, organisations benefit from enlisting crisis communication experts to guide these delicate situations. Crisis communicators must manage not only the stress of senior leadership but also the diverse and often fragmented perception of internal and external stakeholders, including employees, customers, regulators, and the media. Rapid, clear decision-making and concise messaging are essential in these moments when audiences are wide-reaching and expectations are high.

This chapter will cover the core principles for preparing cyber crisis communication effectively, starting with monitoring systems, preparing adaptable crisis messaging, training spokespersons, and running simulations tailored to cyber scenarios. Because while they may appear calm and collected, effective crisis communicators are anything but improvisational – they're well-practised, ready to face multiple possible scenarios and adapt with confidence. After all, communicating effectively in a crisis is essential to maintaining trust and credibility, which is the foundation on which resilience is built.

# Same but Different

The fundamentals of honesty, empathy, and transparency remain at the core of any crisis response, regardless of the nature of the crisis. Yet when a cyber crisis is under way, these principles take on unique dimensions that shape how organisations communicate and respond.

Cyber crises, unlike other crises, are inevitable. They aren't isolated or one-off events; rather, they have the potential to escalate quickly, with incidents in one region impacting operations in another within seconds. The way an organisation communicates during a cyber crisis can mean the difference between rapid recovery and prolonged decline. Honesty, empathy, and transparency aren't just ethical ideals – they are essential tools for handling the complexities of a cyber crisis. Our experience has shown that when organisations embrace these values, they emerge from a crisis with minimum damage, if not an enhanced reputation.

---

**Mini-Case Study: Maersk and NotPetya (2017)**

The NotPetya cyber attack in 2017 brought one of the world's largest shipping companies, Maersk, to a near-standstill, impacting its ability to conduct business for over a week. While Maersk suffered an estimated $300 million loss, their response demonstrated a highly proactive communication approach. Within hours of the breach, they issued a public statement explaining the impact on operations, set up dedicated media hotlines, and provided continuous updates on their recovery process. According to a study by the Ponemon Institute,[1] Maersk's approach was instrumental in preserving trust and preventing longer-term reputational damage. Maersk's rapid and transparent communication became a benchmark for effective crisis communication, underscoring the need for clear, coordinated messaging in the wake of cyber attacks.

**LESSONS LEARNED**

Swift communication and transparency are pivotal in managing perception and minimising reputational damage, particularly for companies with global operations and a broad customer impact.

---

[1] "The 2017 cost of data breach study: Your response plan is key", Ponemon Institute.

## Honesty

Honesty in crisis communication, as defined by Coombs,[2] is the bedrock of credibility. An organisation's trustworthiness hinges on its commitment to providing accurate, complete information. In a crisis, attempting to hide or downplay the truth can lead to irreparable trust damage, amplifying the crisis's impact. A prime example is Facebook's handling of the Cambridge Analytica scandal (2018–2019). When details of data misuse emerged, Facebook had to deal with accusations of concealment and manipulation, a scenario that significantly impacted its public image. A PwC study[3] supports the importance of honesty, revealing that 92% of companies practising transparent communication during a crisis restored their reputation more swiftly than those that tried to conceal the facts.

*"Honesty is not only the quickest path to recovery but the only one."*

In a recent discussion, a senior leader at a prominent PR firm shared her uncompromising stance when managing client crises. "I make it very clear upfront," she noted, "that we're stepping into a 'bullshit-free zone' from here on out." She emphasised that honesty is not only the quickest path to recovery but the only one. "I can't help a client who isn't being truthful with me," she explained. Her approach sets a decisive tone from the outset, creating a space where facts – however uncomfortable – are openly acknowledged. This clarity enables her team to avoid compounding the crisis with misinformation and allows her clients to focus on the steps needed to regain control. By establishing these parameters, she ensures that communication strategies are grounded in reality, empowering clients to move forward with integrity.

## Empathy

Empathy during a crisis fosters a positive perception of an organisation and encourages acceptance of its response. Research by Naomi Eisenberger and others[4] highlights the critical role empathy plays in social interactions, and this becomes even more vital during a data breach. Consider Marriott

---

[2] Coombs, W. T., et al., "Situational crisis communication theory (SCCT) and application in dealing with complex, challenging, and recurring crises", in *Advancing Crisis Communication Effectiveness: Integrating Public Relations Scholarship with Practice*, Routledge, November 2020.

[3] "Crisis communication", PwC's Global Centre for Crisis and Resilience, www.pwc.com.

[4] Meyer, M. L. et al., "Empathy for the social suffering of friends and strangers recruits distinct patterns of brain activation", *Social Cognitive and Affective Neuroscience*, vol. 8, no 4, 2013, p. 446–454.

International's response to its 2020 data breach, during which millions of customers were affected. By proactively reaching out to individuals and offering free monitoring services, Marriott demonstrated genuine concern for its customers' well-being, which helped soften the blow to its reputation.

However, the ripple effects of a breach extend far beyond direct victims. When personally identifiable information (PII) such as employee data is exfiltrated, the indirect impact on their families becomes a pressing concern – especially when health records are compromised. The exposure of such sensitive information raises privacy and confidentiality issues, causing anxiety and distrust. Family members of affected employees might worry about identity theft, discrimination, or misuse of medical information. Organisations must acknowledge these concerns and act with empathy, recognising that a breach can unsettle entire households, not just individuals.

Ben Hockman, a crisis-response and communications expert, underscores this point, stating that in physical security incidents, "Family members are affected too; they need clarity and due care." This applies equally to cyber incidents. Reaching out to families with updates and reassurance alleviates their anxiety and reinforces the support system surrounding each employee, helping them remain focused on their roles.

For example, offering credit monitoring or identity-protection services tailored to family members shows genuine care and foresight. This not only helps mitigate the emotional toll of a breach but also strengthens trust in the organisation.

A Forrester Research[5] report supports this approach, showing that when companies demonstrate empathy during crises, consumer relationships are improved and loyalty is increased by 36%. Internally, empathy-driven communication boosts employee support by 50%, a critical factor in navigating crises that demand collective focus and effort. Recognising and addressing the indirect impacts of a breach ensures that empathy extends beyond immediate stakeholders, fostering resilience and trust across the broader community.

## Transparency

Being transparent means doing more than just sharing facts; it also means ensuring that communication is clear, precise, and accessible. According to the

---

[5] Gill, M., "In times of crisis, empathy fuels loyalty", Forrester Blog, https://www.forrester.com, 18 December 2020.

transparency theory of Fung, Graham, and Weil,[6] transparency builds trust by making complex information understandable for stakeholders. This principle is especially critical in cyber crises, when people using technical jargon can alienate consumers and the wider public. A recent example is Zoom's response to cyber security concerns during the 2021–2022 period. With privacy issues under scrutiny, Zoom took a transparent approach, issuing regular updates and security reports to detail the measures it was taking. Deloitte's analysis[7] underscored the value of transparency. The report showed that being transparent reduces negative effects on company stock by 22% and enhances customer loyalty by 26%. Additionally, 78% of consumers expect companies to be transparent during crises.

## Honesty, Empathy, Transparency: An Integrated Approach

In cyber crisis communication, honesty, empathy, and transparency aren't just guidelines; they are inseparable pillars of a robust communication strategy. Each complements the others, creating a holistic approach that addresses not only the tactical requirements of crisis management but also the human and relational aspects. Removing any of these elements risks compounding the crisis with a communication failure, resulting in an unnecessary reputational crisis on top of the existing cyber incident.

An effective cyber crisis communication strategy integrates these three principles seamlessly. As highlighted in a *Harvard Business Review* article,[8] organisations that weave honesty, empathy, and transparency into their crisis communication approach are 45% more likely to restore their reputation quickly and reduce the long-term negative impact. This comprehensive approach not only manages the immediate crisis but also fortifies relationships with stakeholders, preserving trust and credibility for the future.

# Proactive Communication: Taking Ownership of the Narrative

In any crisis, timing in communication is paramount – not only for providing clarity but for establishing authority as the primary, trusted source of information. Rather than allowing others – be it customers, competitors, media, or even attackers – to shape the narrative, taking ownership of the story early sets the

---

[6] Fung, A., et al., *Full Disclosure: The Perils and Promise of Transparency*, Cambridge University Press, 2007.

[7] "Silence the noise", *Deloitte Review*, Issue 24, January 2019.

[8] Holtom, B. H., et al., "5 Tips for communicating with employees during a crisis", *Harvard Business Review*, 9 July 2020.

organisation as the reliable voice during uncertainty. This proactive approach affirms control, signals preparedness, and reassures stakeholders, reinforcing that the organisation is effectively managing the situation from the outset.

A recent study by Forrester[9] underscores that proactive communication isn't just a good strategy; it's essential for effective crisis management. Organisations that communicate early and openly are often better positioned to maintain trust and resilience. When Equifax delayed disclosure of their 2017 data breach, it led to widespread distrust, heavy financial losses, and significant executive turnover. In contrast, Marriott International's rapid response to a data leak in 2020 provided a model of how quick, direct communication can protect reputation and calm stakeholders.

## Tailoring Messaging by Stakeholder Group

During a cyber crisis, messaging should be tailored to meet the specific needs and expectations of each stakeholder group. For example:

- Internal Teams: Employees need actionable steps to secure data, including instructions for shutting down systems, changing passwords, or avoiding specific applications. Regular updates maintain morale and keep teams focused.

- Regulators and Authorities: Compliance is critical and providing factual and transparent updates helps mitigate regulatory risks. In the 2020 breach of European healthcare provider Fresenius, close coordination with the EU's regulatory bodies minimised compliance complications and streamlined crisis management.

- Customers and Partners: Proactively informing customers and partners reassures them that their data and business interests are protected. A *Harvard Business Review* study on crisis response emphasises that companies communicating early and directly with clients are 45% more likely to maintain customer trust.

## Unique Aspects of Cyber Crisis Communication

Cyber crises present challenges that differ from those of traditional crises. While they share the potential for reputational, financial, and operational impact, cyber crises require specific strategies tailored to the digital and interconnected nature of the threat.

---

[9] Bruce, I., et al., "Don't wait for a crisis to act", Forrester Blog, https://www.forrester.com, 23 July 2024.

■ **Attacker Intelligence and Intentions**

Unlike natural disasters or product recalls, cyber attacks are perpetrated by motivated, organised individuals or groups with in-depth technical knowledge. These attackers are often skilled in exploiting information systems to steal, damage, or compromise data, making it critical for communicators to acknowledge the sophistication of the threat and highlight the organisation's proactive response.

■ **Rapid Propagation**

Cyber threats spread at breakneck speed. In interconnected environments, a single breach can affect multiple systems in seconds, impacting operations, databases, and customer-facing platforms. Communicators must be prepared to respond in real time, often before they have a complete understanding of the impact, to contain reputational damage and maintain trust.

■ **Technical Complexity**

Explaining the origins and implications of cyber incidents to a non-technical audience is challenging. Cyber crises often involve cryptography, system vulnerabilities, and forensic investigations, requiring communicators to simplify complex terms while conveying accountability and expertise.

■ **Anonymity and Attribution Challenges**

Cyber attackers frequently operate from the shadows, with identities hidden behind layers of anonymity. Determining and attributing the source of an attack is difficult and can lead to missteps if handled prematurely. For this reason, entities like France's ANSSI (*Agence nationale de la sécurité des systèmes d'information*)[10] advise against public attribution without solid evidence, as misinterpretation can heighten geopolitical sensitivities.

■ **Heightened Reputational Risk**

The perception of a failure to secure sensitive information can leave lasting damage, especially as the reliance on digital systems grows. For many stakeholders, a cyber breach feels like a violation of trust in a way that is unique to the digital space.

---

[10] Equivalent to the UK's NCSC.

## *Precision*

In a cyber crisis, clear, technically accurate messages are essential to demonstrate control, while transparency about protective and recovery measures builds confidence in the organisation's response.

Cyber crisis communication requires a unique blend of immediacy, technical clarity, and transparency. It's a delicate balance and each word matters, every decision influences trust. Communicating effectively in such a landscape is an art, a craft that blends resilience and empathy to support the organisation through its most challenging days.

## *Strategic Preparedness: Monitoring*

Effective monitoring is more than an asset; it's essential for successful cyber crisis communication and response. By continuously gathering and analysing information – ranging from intrusion attempts and emerging threats to competitor movements and regulatory changes – companies can anticipate crises and react swiftly to mitigate damage.

Organisations that embed intelligence-gathering into their crisis communication strategy not only reduce potential damage but also fortify their resilience. For communication experts and crisis strategists alike, staying informed about risks is the bedrock of managing them. Risk management doesn't come from luck; it requires anticipation, expertise, and a robust intelligence framework.

Information intelligence encompasses the collection, analysis, and application of relevant data to proactively address threats. In the context of cyber crises, intelligence allows organisations to detect vulnerabilities, track trends, identify weak signals, and monitor discussions across social networks and hacker forums. For example, the attack on SolarWinds was first detected through proactive monitoring. FireEye, a cyber security firm, identified suspicious activity patterns in their Orion software, enabling them to alert other users and contain the breach. Similarly, Marriott International's cyber security team detected an attempt to sell stolen data on hacker forums, allowing for a rapid response.

According to CrowdStrike, [11] proactive monitoring on the dark web can provide a crucial advantage, as "actively monitoring the dark web. . . gives cyber criminals less time to exploit your confidential information, preventing further information leaks." Information intelligence is therefore an indispensable

---

[11] Lenaerts-Bergmans, B., "Dark web monitoring", www.crowdstrike.com, 27 April 2023.

part of any cyber security strategy, even though many organisations still face challenges in its implementation.

## Building an Effective Intelligence System

Establishing an intelligence system goes beyond passive observation; it requires hearing and interpreting signals to drive action. An effective monitoring strategy typically includes three pillars:

1 Setting Clear Objectives: Understand the organisation's specific needs and establish priorities to ensure that monitoring aligns with key goals.
2 Leveraging Advanced Technology: Using AI and machine learning to automate data collection and analysis streamlines the monitoring process and allows for more nuanced insights.
3 Assembling a Dedicated Team: Intelligence teams often include cyber security experts, data analysts, communication professionals, and legal advisors, ensuring a well-rounded approach.

Some companies, including Microsoft, have made significant investments in active monitoring. The tech giant's global monitoring centre uses sophisticated algorithms to analyse vast amounts of data, providing security and communications teams with real-time insights. Equinix, a digital infrastructure provider, employs scraping techniques to monitor hacker forums, allowing them to detect and respond to potential threats early. It's the classic game of cat and mouse, with intelligence tipping the scales in favour of the defender.

## From Knowledge to Action

Monitoring alone is insufficient; organisations must use insights from intelligence to guide their crisis communication strategy. Intelligence helps not only cyber security teams but also communicators, allowing them to develop prepared responses, examine potential crisis scenarios, and ensure consistent messaging.

For instance, Leclerc in France recently began positioning itself as a distributor of medical devices through early television ads, likely a decision made as a result of market intelligence. In the context of cyber crisis communication, intelligence aids in crafting clear, structured messages for rapid deployment, training spokespersons to address crises effectively, and ensuring that responses are both credible and timely. Target exemplifies this approach. Faced with increasing cyber threats, the retail corporation integrated intelligence into its communication strategy, preparing response scenarios and pre-approved messages to ensure it would issue coordinated, swift responses.

# The Case for Intelligence in Cyber Crisis Communication

Research reinforces the value of intelligence-driven monitoring. A Deloitte report[12] highlighted the firm's use of information intelligence to enhance detection and response capabilities for clients. IBM[13] adopts a similar approach, observing "new attack vectors and emerging malware" to protect customers effectively. When structured and deployed effectively, intelligence-driven communication is more than just a defence; it's an enabler of resilience and trust.

## Collaborating with IT and Cyber Security Teams for Clear Crisis Communication

Translating technical information from IT and cyber security teams into clear, accessible messages is essential yet challenging for effective cyber crisis management. To reassure and mobilise all parts of an organisation's perception ecosystem, communications teams need to bridge the gap between technical jargon and public understanding. If companies can keep their communications and cyber security departments collaborating well, use simplified language, establish pre-set communication protocols, and consult experts to interpret technical details, they can better manage cyber crises, safeguard their reputation, and maintain stakeholder confidence before, during, and after an incident.

Unlike other crises, cyber incidents often stem from highly technical origins. This technical complexity can make collaboration between communication and IT teams particularly important – and challenging. While every profession has its own specialised language, the distinct nature of IT and cyber security can create barriers. As senior leaders often note, "the CIO is the strangest person on the Executive Committee", and that's a testament to the perceived

---

[12] "Detect & Respond: Staying ahead of cyber threats", Deloitte, www.deloitte.com/cyber, 2021.

[13] Henderson, C., "Bringing threat intelligence and adversary insights to the forefront: X-Force Research Hub", IBM, https://securityintelligence.com/x-force, 3 August 2023.

differences between technical and non-technical departments. These differences are understandable, given that IT has only recently become ubiquitous, while other areas like personnel management and finance have been around for centuries. Despite these challenges, effective communication between technical and non-technical teams remains essential.

The intricate nature of cyber threats does not make a crisis communicator's job easy. For effective communication, technical information from IT and cyber security teams must be translated into language that is clear to all audiences – internal and external. Employees need to know the security steps to follow, understand how the incident might affect their work, and be clear on their responsibilities during the crisis. Customers and partners must be reassured about any impact on their data or services, while the public needs messages that inspire confidence in the organisation's ability to manage the crisis. Additionally, clear communication is necessary to meet regulatory requirements such as the GDPR, which demands transparency in data breach notifications, or ISO 27001, which calls for well-defined communication on security incidents.

A 2016 Deloitte report[14] found that companies coordinating between departments during cyber crises achieved stronger crisis management outcomes. Establishing clear points of contact and consistency in messaging is key. Success requires a collaborative effort, with dedicated training and mutual understanding between Communications and Cyber Security representatives. Regular simulation workshops with a communication focus, alongside ongoing training on cyber trends and threats, are particularly effective. Furthermore, as with other complex topics, using simple terms and relatable metaphors can make it easier for non-technical audiences to understand. Thoughtful, ethical storytelling helps rally the necessary support from internal and external audiences.

Everyone is responsible for contributing to the resolution of a cyber crisis; not just the IT department. Think of it like a ship taking on water – if only the deckhands were bailing out the hull, it wouldn't be enough. Similarly, during a cyber attack, all hands must be on deck.

---

[14] "Cyber crisis management: readiness, response and recovery", Deloitte, https://www2. deloitte.com/content/dam/Deloitte/ch/Documents/audit/ch-en-cyber-crisis-management. pdf, 21 April 2016.

## Common Pitfalls in Cyber Crisis Communication

Even the best-prepared organisations can fall into communication traps during a crisis. Common pitfalls include:

■ Overuse of Technical Jargon: Complex language alienates non-technical stakeholders. During the Target data breach in 2013, jargon-heavy responses created confusion and frustration among customers and investors.

■ Delayed Communication: Holding back information can spark rumours. A 2021 PwC report[15] found that customer trust dropped by 30% more for companies that delayed responses during cyber incidents.

■ Inconsistent Updates: Maintaining consistent messaging prevents misinformation and speculation. Facebook's response to the Cambridge Analytica scandal was criticised for inconsistent updates, which intensified scrutiny and eroded trust.

Solution: Avoid these pitfalls by crafting clear, jargon-free messaging, ensuring timely communication, and updating regularly to keep stakeholders informed and aligned.

# The Communication Kit

Once monitoring is established and a strong link exists between Communication and Cyber Security teams, it's essential to prepare a comprehensive cyber crisis communication kit. The goal is to be prepared with adaptable resources that can be quickly deployed and combined when needed – knowing that no scenario will fit neatly into what's been anticipated. Having these kits ready helps avoid improvising under pressure and facilitates more effective responses. So, what should this cyber crisis communication "kit" contain?

■ Key Messages: Pre-written messages for various cyber crisis scenarios, tailored to specific audiences (internal, customers, media, regulators), and aligned with the organisation's digital risk matrix.

■ Press Release Templates: Detailed templates for press releases that can be quickly customised and distributed.

■ Q&A (Questions and Answers): An FAQ-style document anticipating potential questions and providing prepared responses and clear procedures.

---

[15] "Four fault lines show a fracturing among global consumers", Report: PwC's Global Consumer Insights Pulse March 2021, www.pwc.com.au, June 2021.

■ Media Contacts: A list of media contacts, including journalists specialising in cyber security. Regularly updating and maintaining these contacts ensures quick mobilisation if a crisis arises.

■ Communication Channels: A plan detailing the channels to be used to reach different stakeholders – email, social media, press conferences, secure messaging, etc.

■ Technical Documentation: Simplified explanations of incident types and mitigation measures, making technical aspects accessible to non-expert audiences.

■ Visual Resources: Infographics, diagrams, and other visual aids to help communicate technical aspects clearly to all stakeholders.

Once your crisis communication kit is prepared, the next step is regular testing and updating. As risks evolve, so should your kit. Ensuring consistency with your digital risk matrix is key, which requires ongoing collaboration between Communications and Cyber Security teams. An annual immersive crisis exercise focused on cyber crisis communication for senior management can help ensure readiness. These exercises reveal the value of predefined kits and back-up communication channels that maintain contact, relay instructions, and share updates, even if primary systems are compromised.

## Playing in Position

Another critical element of preparation is defining clear spokesperson roles before a crisis occurs. Understanding who speaks, what they say, and who steps in as backup is essential for clear, coordinated communication. The key spokesperson roles in a cyber crisis include:

■ Corporate Spokesperson: A senior management member who communicates the impact on the organisation and strategic response measures. Generally, avoid placing the CEO, Chairman, or Communications Director in this role, as it can lead to unintended media focus.

■ Technical Spokesperson: An expert in cyber security who can explain the technical details of the crisis in a way that's accessible to executives and board members.

■ Legal Spokesperson: A legal representative ensuring that all communications meet regulatory and legal requirements.

■ Customer Service Spokesperson: A customer service lead who disseminates crisis messages to address customer and partner concerns directly.

These roles should be practised regularly in the form of media training and realistic simulation exercises. Media training involves placing spokespersons in front of a camera or microphone and carrying out mock press interviews with a "journalist" (ideally someone with journalistic experience). Crisis-simulation exercises put the entire team through a plausible scenario, testing their knowledge, roles, kit quality, and communication channels. Each exercise produces an observation report, offering insights on how to refine communication kits as part of a continuous improvement process.

This approach is widely supported by industry experts. As noted by Wavestone, "When a cyber security incident occurs, a well-defined crisis communication strategy is essential. Such a strategy can make all the difference in ensuring the company remains a trusted partner, retailer, or supplier."[16]

# The Friendly Match

Testing is the ultimate measure of a crisis communication plan's effectiveness. Ensuring readiness involves structured roles, reliable communication channels, and routine simulations that mimic real-world conditions.

## Applying the OODA Loop in Cyber Crisis Communication

In high-stakes situations like cyber incidents, using the OODA loop – Observe, Orient, Decide, Act – can structure communication effectively:

- Observe: Detect changes in the situation, such as escalation of the threat or media interest.
- Orient: Assess impact and alignment of responses with your communication objectives.
- Decide: Choose the most appropriate message and channel for the stage of the crisis.
- Act: Execute communication plans, ensuring timely updates and adapting as needed.

For instance, during the Capital One data breach in 2019, executives used an iterative approach to provide timely updates, using the OODA model to assess and adjust messaging in real time. Studies from the National Communication Association indicate that frameworks like OODA significantly improve response cohesion and stakeholder confidence in crises.

---

[16] "How to create Crisis Comms that will take your crisis response to the next level", Wavestone, https://wwa.wavestone.com, 2024.

## The Importance of the Pilot and Navigator

Effective crisis communication relies on strong coordination, often best achieved through a "pilot" and "navigator" model. This approach provides a structured yet flexible method for managing a crisis, minimising immediate damage, and preserving organisational trust in the long term. Based on experience, we find that organisations unfamiliar with this model gain significant insights when introduced to it, recognising its value in structured, high-pressure situations.

In this model:

■ The pilot is often a senior communicator or experienced leader responsible for overseeing the communication strategy. They ensure messages remain aligned with the company's goals and are delivered consistently. For example, during the 2014 Sony Pictures cyber attack, then co-president Amy Pascal guided the firm's communication response, demonstrating the crucial role of a strong leader in managing a crisis narrative.

■ The navigator devises and adapts communication strategies based on potential developments, anticipating scenarios and preparing responses for likely outcomes. They work several steps ahead, analysing unfolding situations and adjusting tactics accordingly.

A helpful analogy compares the pilot-navigator duo to a rally car team. The pilot manages steering, acceleration, and immediate reactions, while the navigator warns of upcoming twists and turns, guiding the pilot to avoid unseen obstacles. Without the navigator's insights, the pilot's journey would be slower and riskier. This coordination ensures that the crisis is brought to a faster and safer conclusion. While this chapter emphasises crisis communication, the pilot-navigator concept can be applied to many other functions like IT, cyber security, finance, and HR, in the context of crisis management.

Testing this pilot-navigator approach during cyber crisis simulations is highly recommended. Set up a scenario with senior leadership divided into two groups: pilots and navigators. Place the CEO and Communications Director with the pilots, while the Risk Director and a cyber security manager take on navigator roles. This exercise often results in "aha!" moments, when participants realise the value of the pilot-navigator distinction, as illustrated by feedback from an industry CEO who remarked, "I'd never considered this pilot-navigator distinction – it makes so much sense!" Such moments demonstrate the effectiveness of the approach in preparation for a real crisis.

# Crisis Communication Channels

Establishing specific crisis communication channels and testing them beforehand is essential. Surprisingly, many organisations lack dedicated crisis channels when they begin a training programme. Relying on regular telephony and email systems is risky, as these channels may be compromised during an attack. A proper communication structure, like the PACE model (Primary, Alternate, Contingency, and Emergency), safeguards communication continuity, ensuring that messaging remains operational even if standard channels fail.

In a crisis, these channels transmit critical information from a source to one or more receivers, connecting the organisation to its audiences. Various channels offer different advantages in terms of cost, security, and public accessibility. Examples include:

- Garmin's 2020 ransomware attack: Garmin used a combination of its website, social media, and direct email updates to inform customers about recovery progress.

- Colonial Pipeline's 2021 ransomware attack: Colonial Pipeline provided updates through press releases, social media, and a dedicated hotline.

- Norsk Hydro's 2019 cyber incident: The company used Facebook to share images of system repairs and YouTube for press briefings led by its CFO, an unconventional yet effective choice.

## Communication Timelines

Determining the cadence of communication is crucial to maintaining trust during a prolonged cyber incident. Research from Forrester[17] shows that communicating every two hours during the initial phase of a crisis, then scaling to four-hour intervals as recovery stabilises, can mitigate stakeholder anxiety.

- Initial Phase: Every one to two hours, address immediate impacts and outline actions.

- Stabilisation Phase: Every four hours, provide progress updates, even if there's minimal new information.

- Resolution Phase: Release updates as necessary, focusing on post-crisis improvements and preventative measures. In the case of the WannaCry incident that affected the NHS in 2017, structured updates every few

---

[17]  Bruce, I., et al., Forrester Blog, op. cit.

hours helped reassure the public and showed transparency despite an evolving crisis.

## Internal Crisis Communication Channels

Effective internal channels keep employees informed and aligned with crisis management efforts. Common channels include:

■ Internal Emails and Emergency Newsletters: For direct communication with employees, providing real-time updates on the situation.

■ Intranet and Collaboration Platforms: Essential for internal coordination. Microsoft, for example, used Teams to coordinate its response to a vulnerability in 2021. Google Workspace, Slack, and similar tools also serve this purpose well.

## External Crisis Communication Channels

Effective external channels maintain trust and control the crisis narrative. Key options include:

■ Emergency Websites and Press Releases: Critical for publishing official updates.

■ Social Media: Provides fast message distribution and serves as a fallback if internal systems are isolated from the internet. For instance, when Facebook experienced an outage in 2021, it used Twitter to update users.

■ Hotlines and Emergency Numbers: Enable direct information access for customers.

■ SMS and Push Notifications: Ideal for conveying urgent information rapidly.

## Choosing the Right Communication Channels

When applying the PACE model, consider security, usability, and user familiarity. For example, in companies with less tech-savvy employees, a public-facing, user-friendly collaboration tool may be better suited than an ultra-secure but unfamiliar platform. Additionally, inform stakeholders in advance about which channels you will use in a crisis. This reduces the risk of misinformation from fake channels, as seen in the Equifax case, where attackers set up a fraudulent rescue site.

Testing channels during "peacetime" is crucial. Simulations, tabletop exercises, and regular audits validate each channel's reliability and equip leaders with familiarity in real-life situations, minimising errors and stress when a real crisis unfolds.

## Organising Cyber Crisis Communication Exercises

In Chapter 5, we discussed the organisation of cyber crisis exercises for Executive Management, covering all business lines represented at that level, including communications. When focusing specifically on cyber crisis communication, the approach is similar but with a narrower scope. These exercises are typically shorter (around 2.5 hours), require fewer resources from the immersion team, and can serve as an effective, lower-cost way to initiate continuous improvement processes with senior management.

That said, organising an effective cyber crisis communication exercise requires serious preparation, finesse in facilitation, and comprehensive post-exercise follow-up to solidify learned reflexes and improve the organisation's communication approach in future crises.

# From Preparation to Evaluation

## Preparation Phase: Defining Goals and Scenarios

In the preparation phase, it's crucial to align the exercise objectives with the company's current and upcoming projects, along with highest-priority cyber risks. This alignment ensures relevance and engagement, especially with senior management, and helps demonstrate that the exercise is tailored to the organisation's real-world challenges.

- Define Objectives Clearly: Take time to discuss and define the training goals with both the Executive Management participants and the cyber security lead (such as the CIO or CISO). Regular check-ins are beneficial to capture changes in the organisation's risk profile or strategic direction, ensuring each exercise feels relevant even if conducted annually.

- Select Participants Thoughtfully: Typically, extended General Management is included (about ten people), and it's important to involve the Group Communications Director if you're training a subsidiary. This allows all key voices to be part of the scenario, fostering an environment where communication strategies can be tested in all areas of leadership.

■ Design Realistic Scenarios: Use a sequence of worst-case situations based on plausible threats to the organisation. A scenario that is grounded and aligned with the latest insights from your internal partners will resonate strongly with participants, helping them see the relevance of the exercise.

## Execution Phase: Flexibility and Live Coaching

During execution, follow your prepared scenario, but be responsive to participant reactions. Leaders may have unanticipated insights or responses, and adapting on the spot maintains realism and engagement. This phase is the perfect time to employ the PACE (Primary, Alternate, Contingency, and Emergency) trainer model, which involves keeping backup plans ready to handle unforeseen developments without deviating from training goals.

■ Use Multi-Media Stimuli: Leverage different media to create a rich, immersive experience. Photos, videos, simulated press releases, fake news broadcasts, phone calls, and even sound effects can deepen engagement, representing the real-world noise and distractions of a crisis. This variety keeps the pace dynamic and participants focused.

■ Emphasise Real-Time Coaching: Rather than merely delivering information, add a layer of live coaching to support participants as they navigate the scenario. This personalised guidance helps them internalise key communication techniques and recognise the flexibility needed to handle evolving situations. As one CEO remarked, "It was so well-prepared and supervised that we had space to improvise." This feedback underscores the value of balancing structure with room for adaptive learning.

## Evaluation Phase: Constructive Feedback and Continuous Improvement

In the final phase, the goal is to capture observations and provide constructive feedback to Executive Management, consolidating insights gathered during the exercise.

■ Detailed Observation Report: Use data collected during the exercise and notes from the live training to develop an observation report. This report should include specific recommendations for strengthening the organisation's crisis communication strategies.

■ Balancing Feedback: Presenting feedback to senior leaders requires a blend of expertise and tact. The focus should be on development rather than critique, emphasising the organisation's strengths and practical steps to improve in the future. Successful feedback sessions often rely on qualities like honesty, empathy and transparency, as well as a good dose of benevolence and expertise.

▪ Support a Cycle of Continuous Improvement: After delivering feedback, remain available for follow-up, ideally within six to twelve months. This continuity allows you to support ongoing improvements, helping the organisation adapt to evolving threats and reflect on the recommended measures it has implemented.

This approach, combining meticulous preparation, responsive facilitation, and supportive evaluation, ensures that crisis communication exercises are impactful and well received by leadership, reinforcing the organisation's resilience in cyber crises.

# Building an Adaptive Communication Framework

A robust communication framework is the key to handling the unpredictable nature of cyber crises effectively. Preparing a framework that is both adaptable and structured enables organisations to manage real-time developments, protect their reputation and maintain stakeholder trust. This section explores key elements to consider when building such a framework.

## 1. Establish Core Principles

At the foundation of any effective framework are the core principles of communication: honesty, empathy, and transparency. These values guide all responses, helping maintain consistency and integrity across channels. For instance, in any communication following a data breach, leaders should openly acknowledge the issue, express empathy for those affected, and provide transparent updates about the next steps.

## 2. Define Communication Objectives

Clear objectives help focus the message and ensure that communication aligns with the organisation's crisis-response goals. In a cyber crisis, these objectives might include:

▪ Informing Stakeholders: Deliver timely and relevant updates to those impacted.

■ Reinforcing Trust: Maintain trust by being transparent about actions taken.

■ Mitigating Misinformation: Control the narrative to prevent the spread of rumours or misinformation.

Setting these objectives before the impact of a crisis is felt enables rapid and targeted communication and ensures that messages serve their intended purpose.

## 3. Develop and Pre-Test Message Templates

Creating and testing message templates in advance helps streamline communication when time is critical. These templates should cover a range of scenarios, such as data breaches, ransomware attacks, or system outages, and be tailored to various audiences, including employees, customers, regulators, and the media.

**Example Template Components:**

■ Initial Notification: Brief, factual statement acknowledging the incident.

■ Updates: Clear, regular updates on the status of the crisis and actions being taken.

■ Resolution: Communication on the closure of the incident and any long-term preventive measures.

Pre-testing these templates through crisis simulation exercises ensures they are effective under pressure, allowing adjustments before a real crisis occurs.

## 4. Establish Crisis Communication Channels

Identify primary, alternate, contingency, and emergency (PACE) communication channels to serve as reliable and accessible communication pathways. It's essential to select channels based on the audience, security level, and accessibility in a crisis:

■ Primary Channels: Internal email, collaboration platforms, and intranet sites.

■ Alternate Channels: Secure messaging apps or internal newsletters.

■ Contingency Channels: Public social media accounts or press releases. Emergency Channels: Hotline numbers, SMS alerts, or pre-established media contact networks.

Testing these channels regularly ensures they function as planned, and training staff to use them during simulations prepares them to communicate effectively if regular channels are compromised.

## 5. Designate Spokespersons and Roles

Each spokesperson should have a clearly defined role, tailored to their expertise and audience, to prevent overlapping messages and confusion. Common roles include:

- Corporate Spokesperson: Delivers updates on overall impact and high-level strategy.
- Technical Spokesperson: Provides clear, simplified explanations of technical aspects.
- Customer Service Spokesperson: Handles direct communication with affected customers.
- Legal Spokesperson: Ensures that communications comply with regulatory requirements.

Training and role-play simulations enhance spokesperson confidence, enabling them to stay composed and articulate under pressure.

## 6. Integrate with Crisis-Response Teams

Effective communication requires seamless coordination with the crisis-response team, including cyber security, IT, legal, and public relations. This integration ensures that messaging aligns with technical response efforts, creating a cohesive response that reinforces trust.

Regular meetings and shared platforms between teams enable real-time information sharing, ensuring that communicators are fully informed and able to adapt messages as the situation unfolds.

## 7. Implement Feedback Loops

After each communication exercise or real crisis, gather feedback to refine the framework. This feedback, along with quantitative metrics (such as message reach and stakeholder sentiment), supports continuous improvement, keeping the framework agile and responsive to evolving needs.

## Final Thoughts

Building an adaptive communication framework is about creating a flexible yet structured system that can respond to unforeseen challenges with clarity and resilience. By preparing in advance, organisations can handle crises with greater agility, ensuring that each message protects both their reputation and stakeholder trust.

---

**Case Study: Saint-Gobain, Global Ransomware Attack, 2017**

**The Facts**
In June 2017, Saint-Gobain, a prominent player in the global building materials industry, faced a significant cyber incident when it became a target of the NotPetya ransomware attack. The attack disrupted Saint-Gobain's operations, ultimately costing the group around €80 million in operating income. Beyond this immediate disruption to its business and finances, a crucial aspect of managing this sensitive situation was how Saint-Gobain communicated.

**Focus on Crisis Communication**
From the start of the attack, Saint-Gobain's management team recognised the importance of transparent, proactive communication. The initial response was to promptly inform employees and internal stakeholders. An emergency email alerted all staff to the potential risks and outlined necessary steps, such as immediately shutting down IT systems to contain the spread of the malware.

Externally, the group quickly issued a public statement confirming that it was facing a major cyber attack. By making that early public statement, Saint-Gobain aimed to control the narrative, preempt speculation, and minimise panic. The organisation committed to regular public updates for customers, suppliers, investors, and the media, to maintain confidence and trust throughout the crisis.

**Communication Channels Employed**
Saint-Gobain leveraged a range of communication channels to keep stakeholders informed:

■ Internal e-Newsletters: Regular updates were provided to employees, including those not directly involved in crisis response, through e-newsletters and internal meetings. These communications helped sustain employee confidence and rally collective support to overcome the crisis.
■ Website and Social Media: Saint-Gobain used its website and social media accounts to issue updates, reaching a broad audience quickly and effectively.
■ Press Releases: Regular press releases provided information on the progress of recovery efforts, operational impacts, and projected timelines for resuming normal operations (though sharing expected recovery timelines carries certain risks).

▶

- Conference Calls and Webinars: Saint-Gobain organised conference calls for investors and financial analysts to deliver real-time updates and address questions. Similarly, webinars were held for the Group's customers and partners to provide reassurance on business continuity.
- Dedicated Call Centre: A specialised call centre was established to handle customer inquiries and concerns, with additional customer service support staff providing personalised assistance and alternative solutions where possible.
- Coordination with Authorities: Saint-Gobain collaborated closely with national and international cyber security authorities, including ANSSI in France. This cooperation allowed the company to align its efforts with expert guidance, strengthening its cyber security posture and optimising its recovery.

Once the attack had been contained, Saint-Gobain focused on recovery and emphasised the improvements made to security systems and protocols. Management conveyed a strong message regarding the importance of cyber security, reportedly stating, "Anyone who does not agree with the cyber security policy has no place in the Group." This message underscored the significance of cyber security in the company's future strategy and illustrated just how severely the crisis had impacted its organisational culture.

### LESSONS LEARNED

The 2017 ransomware attack posed a substantial operational, technological, and financial challenge for Saint-Gobain. In terms of crisis communication, the organisation demonstrated the importance of the following principles:

- Honesty, Empathy, and Transparency: Saint-Gobain adhered to the key principles of crisis communication, providing clear and truthful updates to sustain stakeholder trust.
- Proactive and Multi-Channel Approach: The company's prompt, multi-channel communication strategy helped to control the narrative, address stakeholder concerns, and maintain confidence throughout the crisis.
- Financial Transparency: Saint-Gobain notably became one of the first companies to disclose the major financial impact of a cyber attack on its operations, setting a significant precedent in corporate crisis communication.

The Saint-Gobain case remains a powerful example of how effective a transparent, proactive communication strategy can be during a cyber crisis and demonstrates the company's commitment to honesty and transparency even in challenging circumstances.

Their approach echoes the words of Warren Buffett: "It takes 20 years to build a reputation and five minutes to ruin it. If you think about that, you'll do things differently." Saint-Gobain thought differently, leveraging the crisis to strengthen trust and align internal teams.

This case reminds us that clear messaging and preparation aren't just about survival, they're about seizing the chance to emerge stronger. The following key takeaways provide a blueprint for developing communication strategies that can withstand the complexities of modern cyber crises.

**Key Takeaways**

In Chapter 6, we explored the critical role of communication in effectively managing cyber crises. Recognising that cyber incidents present unique challenges – including rapid propagation, technical complexity, and intentional attacks – this chapter emphasised the necessity of proactive, transparent, and empathetic communication strategies. By integrating core principles, collaborating with all departments and preparing through simulations and frameworks, companies can navigate cyber crises and stay more resilient. Below are the essential insights and actionable points from this chapter:

- **Honesty, Empathy, and Transparency Are Essential Pillars**
  Maintaining these core principles helps sustain trust and credibility, enabling organisations to minimise reputational damage during a cyber crisis.

- **Proactive Communication Is Vital for Managing Perception**
  Taking control of the narrative early prevents misinformation and reassures stakeholders that the situation is being effectively managed.

- **Tailor Messaging to Different Stakeholders**
  Customised communication for employees, regulators, customers, and partners ensures that each group's specific concerns are addressed appropriately.

- **Establish Dedicated Communication Channels and Test Them Regularly**
  Separate, reliable channels for crisis communication are crucial; regular testing ensures they function correctly when needed most.

- **Coordination with IT and Cyber security Teams Is Crucial**
  Close collaboration allows for accurate, accessible messaging that bridges the gap between technical details and stakeholder understanding.

- **Use Monitoring and Intelligence to Anticipate and Respond Quickly**
  Proactive monitoring enables early detection of threats and informs timely adjustments to communication strategies.

▶

■ **The Importance of Regular Cyber Crisis Exercises**
Simulations and drills enhance preparedness, revealing gaps in strategies and building confidence among crisis responders.

■ **Create and Maintain a Crisis Communication Kit**
A prepared kit with resources like message templates and contact lists enables swift, organised responses without the need for improvisation.

■ **Cyber Crises Require Unique Communication Approaches**
The intentional and technical nature of cyber attacks demands precise, flexible communication strategies distinct from traditional crisis responses.

■ **Implement a Continuous Improvement Process**
Post-crisis evaluations and feedback loops refine communication plans, strengthening resilience against future incidents.

# CONCLUSION

"Cybersecurity is always too much, until the day you need it."

– William H. Webster

Muscle memory is an extraordinary thing. When your brain switches into autopilot, the reflexes you've built through consistent practice take over. Think about when you drive. You instinctively know how to change gears, brake, or steer without making conscious decisions. After years of driving, these actions become second nature because you've practised them so many times that they've become automatic, even when you're not actively thinking about them.

Top athletes often say that their "real training starts where everyone else's stops". This mentality is shared by professionals in high-risk fields – whether soldiers, emergency responders, or business leaders who understand that success depends on how you respond when it matters most. And when it comes to cyber crises, the same principle applies. As the CEO of Maersk said after the 2017 cyber attack: "We can't be average, we've got to be the best we can."

For businesses, this drive for excellence isn't just an aspiration; it's essential for survival. As we remind our clients: "Success also means getting back up faster than your competitors." Cyber resilience is no longer just a strategic advantage, it's a necessity.

As Alex Honnold, the first climber to free-solo El Capitan, said: "If you don't think about what could go wrong, you might end up in a mess you haven't prepared for." The same applies to business continuity: without careful planning and preparation, even a small misstep can have serious consequences.

In short, businesses need to be prepared for a cyber attack to maintain their competitive edge and ensure long-term survival. As you finish reading this book, you now have the tools to begin making real progress. Whether you're leading a large enterprise or working within a smaller organisation, the keys to cyber resilience are within reach.

Cyber crime continues to grow, with little risk for perpetrators and immense rewards. Unless we change how we approach digital transformation, the attack surface will only continue to expand. Cyber security is, therefore, the most vital ally any organisation can have in fulfilling its digital ambitions.

It's time to take action. Start investing in your training, your people, and your future. The risks are real, and the consequences of inaction are too great to ignore.

Your journey to cyber resilience starts now. Let's make it count!

# The Cyber Crisis Pocket Guide

## POCKET GUIDE 1: RANSOMWARE ATTACK

Ransomware attacks are a serious and growing threat to organisations of all sizes and industries. Understanding ransomware and how it works is essential for effective prevention and response. Failing to act proactively can lead to severe data loss, financial strain, and reputational damage.

## 1. Overview

■ **What Is Ransomware?**

Ransomware is malicious software that encrypts files, demanding payment for their release. It's a highly disruptive form of attack that can cripple an organisation's operations.

■ **How Does Ransomeware Spread?**

Typically delivered through phishing emails, malicious websites, or exploiting software vulnerabilities, ransomware can spread quickly if defences are weak.

## 2. Attack Life Cycle

■ **Infiltration**

The attack often begins with a phishing email or an exploited software vulnerability, introducing malicious code into the system.

■ **Encryption**

The ransomware encrypts files using strong encryption algorithms, rendering them inaccessible without the decryption key.

■ **Ransom Note**

Once encryption is complete, attackers display a ransom note demanding payment for the decryption key and threatening data release or deletion.

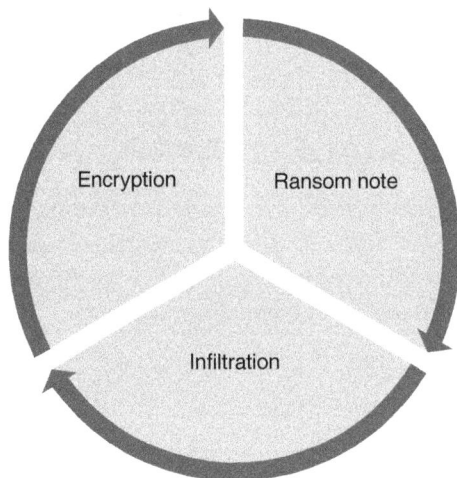

## 3. Consequences of a Ransomware Attack

■ **Data Loss**

Files become inaccessible, potentially causing the permanent loss of critical business data.

■ **Financial Impact**

Ransom payments can be large, and recovery costs (including system downtime, data recovery, and legal fees) can easily exceed the ransom itself.

■ **Reputation Damage**

Public trust may be significantly eroded due to data breaches, extended downtime, or perceived negligence in security practices.

## 4. Prevention Strategies

■ **Employee Training**

Ensure employees are trained to identify and report phishing attempts quickly. Encourage scepticism toward unsolicited email attachments or links.

■ **Regular Backups**

Regularly back up critical data to offline or secure cloud storage. Test restore capabilities regularly to ensure quick recovery in case of an attack.

■ **Software Updates**

Apply patches and updates promptly to operating systems and software, closing known vulnerabilities and reducing the chance of exploitation.

■ **Network Segmentation**

Consider segmenting critical systems and data into separate networks to limit the potential spread of an attack.

■ **Multi-Factor Authentication (MFA)**

Enforce MFA for access to critical systems, making it more difficult for attackers to gain unauthorised access.

## 5. Incident Response Plan

■ **Isolation**

Immediately isolate infected systems to prevent the ransomware from spreading. Disconnect affected devices from the network and secure backup systems.

■ **Incident Reporting**

Establish quick, clear reporting channels for employees to report suspicious activity. A timely response can minimise damage.

■ **Communication**

Transparently communicate with stakeholders – employees, customers, and partners – about the incident and ongoing efforts to resolve it. Keep everyone informed throughout the response.

## 6. Payment Considerations

■ **Be Cautious**

Paying the ransom doesn't guarantee data recovery. In many cases, organisations still fail to regain full access to their data or systems.

■ **Legal and Ethical Considerations**

Evaluate the legal and ethical implications before considering ransom payment. Some jurisdictions may prohibit or penalise ransom payments, and funding criminal activity may have broader societal consequences.

## 7. Post-Incident Recovery

### ▣ Forensic Analysis

Conduct a thorough forensic investigation to understand the origin of the attack, the exploited vulnerabilities, and how to prevent similar attacks in the future. This will guide improvements in security posture.

### ▣ Enhanced Security Measures

Based on lessons learned, strengthen cyber security measures across the organisation. Implement multi-layered security, including real-time monitoring, endpoint detection, and regular vulnerability assessments.

Ransomware attacks demand proactive, comprehensive action. Educating employees, implementing preventive measures, and having a clear response plan are essential for minimising the impact of a ransomware attack. Cyber resilience is a critical part of your organisation's overall strategy. Start preparing now to defend against the growing threat of ransomware.

# POCKET GUIDE 2: PRESIDENT FRAUD AWARENESS

President fraud, also known as CEO fraud or Business Email Compromise (BEC), is a deceptive tactic targeting organisations. Understanding the key characteristics and preventive measures is crucial for safeguarding against this type of fraud.

## 1. Overview

### ▣ What Is President Fraud?

President fraud involves impersonating a high-ranking executive, often the CEO or president, to manipulate employees into unauthorised actions, such as fund transfers or disclosing sensitive information.

### ▣ Rising Threat

President fraud has been rising steadily, with attackers becoming more sophisticated in their tactics. Organisations that fail to act may face significant financial and reputational damage.

# 2. Tactics Used

### ■ Email Spoofing

Attackers spoof emails to appear as if they are coming from a legitimate executive. This often involves meticulous research to mimic communication style and details.

### ■ Urgency and Authority

Impersonators exploit urgency, requesting immediate action. They convey a sense of authority to minimise scepticism and pressure employees into compliance.

# 3. Common Scenarios

### ■ Financial Requests

Fraudulent emails may request fund transfers or sensitive financial information, targeting employees involved in financial transactions.

### ■ Sensitive Data Requests

Impersonators may seek confidential data, such as employee records or customer information, often posing as executives requesting urgent access.

# 4. Consequences

### ■ Financial Loss

Organisations may incur substantial financial losses from unauthorised transactions or data breaches.

### ■ Reputation Damage

Loss of trust from customers, partners, and employees can result from falling victim to president fraud.

### ■ Operational Disruption

Unauthorised actions may disrupt business operations, causing delays and resource drain.

# 5. Prevention Strategies

### ■ Email Authentication

Implement email authentication protocols (SPF, DKIM, DMARC) to detect and prevent email spoofing. Regularly monitor and update these protocols to maintain effectiveness.

■ **Verification Protocols**

Establish clear verification protocols for financial transactions or sensitive data requests. Encourage a culture of double-checking with executives via known communication channels.

■ **Employee Training**

Ensure employees undergo regular training to identify and report phishing attempts and fraudulent emails immediately. Emphasise the importance of verifying unusual requests, especially those involving financial transactions.

# 6. Incident Response Plan

■ **Verification Procedures**

Immediately verify any unusual requests by calling or using an alternative communication method with the executive making the request. Never reply to the same email thread.

■ **Incident Reporting**

Establish a clear reporting mechanism for suspected president fraud incidents. Ensure prompt investigation and a swift response.

■ **Communication**

Communicate transparently with employees about the threat of president fraud, reinforcing the importance of vigilance and encouraging the reporting of suspicious activities.

# 7. Legal Considerations

■ **Legal Reporting**

Report president fraud incidents to law enforcement and comply with all legal reporting requirements. Cooperate fully with investigations, ensuring that all necessary documentation is provided for prosecuting offenders. Prompt reporting can also help identify patterns across industries.

■ **Liability Assessment**

Work with legal experts to assess potential exposure to fines, lawsuits, or regulatory scrutiny in the event of financial losses due to president fraud. Collaborate with your legal team to determine your responsibilities and any legal repercussions, ensuring that your organisation is prepared to navigate the aftermath of a fraud incident.

## 8. Post-Incident Review

### ■ Forensic Analysis

Conduct a thorough forensic investigation to understand the tactics used during the fraud attempt. Identify vulnerabilities in communication and transaction processes that allowed the fraud to occur. This will guide improvements to your security infrastructure and response strategies.

### ■ Training Enhancement

Following each incident, update training materials to include recent examples of attacks. Ensure that employees are kept aware of new tactics and trends in fraud. Regularly reinforce these lessons in training sessions and ensure employees know how to handle similar situations in the future.

President fraud is a sophisticated threat that demands a multifaceted defence. Employing robust prevention measures, continuous training, and a well-defined response plan are essential for mitigating the risks associated with president fraud.

# POCKET GUIDE 3: OPERATIONAL TECHNOLOGY (OT) INSIDER THREAT

Insider threats within Operational Technology (OT) environments pose significant risks to critical infrastructure and industrial operations. Understanding various OT insider threat scenarios is crucial for preemptive detection and mitigation strategies.

## 1. Types of OT Insider Threats

### ■ Unauthorised Access

**Scenario:** Employees or contractors accessing OT systems beyond their authorised privileges.

**Risk:** Potential manipulation or disruption of critical industrial processes, leading to significant operational downtime and safety risks. Unauthorised access to industrial systems can cause costly disruptions and safety incidents, especially in sectors like energy and manufacturing.

■ **Data Theft or Sabotage**

**Scenario:** Malicious insiders stealing sensitive OT data or intentionally causing damage.

**Risk:** Loss of proprietary technology, operational disruption, and long-term financial and competitive damage. Data theft can result in intellectual property loss, while sabotage may cause costly operational downtime and permanent damage to equipment.

■ **Vendor/Supplier Compromise**

**Scenario:** Insider threat originating from third-party vendors or suppliers with access to OT systems.

**Risk:** Compromising OT systems through manipulated or compromised vendor channels, allowing attackers to bypass security. Vendors with privileged access can provide the ideal vector for malware introduction or espionage.

■ **Malware Introduction**

**Scenario:** Insiders introducing malware into OT networks intentionally or inadvertently.

**Risk:** Disruption of industrial processes, equipment malfunction, or system downtime, potentially causing millions in damages. Malware can lead to equipment failures, halted production lines, and may even cause critical safety incidents.

## 2. Indicators of OT Insider Threats

■ **Unusual Network Activity**

Anomalies in network traffic, such as unauthorised connections or large data transfers, suggest suspicious insider activity. Examples include unauthorised remote connections to OT networks or unusually high data flows outside normal operating patterns.

■ **Abnormal User Behaviour**

Unauthorised access attempts, unusual login times, or irregular access patterns raise alarms for potential insider threats. Pay attention to employees logging in at odd hours or accessing systems they typically don't interact with.

### ■ Data Exfiltration Attempts

Suspicious attempts to extract or transfer sensitive OT data, particularly during off-hours, should be flagged. This can include unusual file transfers or data being sent to unapproved external devices or locations.

### ■ Changes in System Configuration

Unauthorised modifications or alterations in OT system settings may signal an insider attempting to disrupt or compromise operations. This includes changes to safety-critical systems or tampering with alarms, sensors, or control protocols.

## 3. Mitigation Strategies

### ■ Role-Based Access Control (RBAC) )

Implement RBAC to limit access based on job roles and responsibilities. Ensure access rights are regularly reviewed and adjusted to reflect role changes. Access should be granted based on necessity, and systems should be configured to restrict access to critical functions unless absolutely required.

### ■ Continuous Monitoring

Use real-time monitoring tools like SIEM (Security Information and Event Management) or UEBA (User and Entity Behaviour Analytics) to detect anomalies and deviations from normal operational patterns. Continuously track user activity and network traffic for signs of insider activity.

### ■ Employee Awareness Programmes

Regularly train employees on OT security practices and insider threat awareness. Incorporate real-world scenarios into training to improve recognition and response. Employees should know how to report suspicious behaviour and understand the specific threats that exist in OT environments.

### ■ Behavioural Analytics

Leverage behavioural analytics to track user activity within OT systems, helping identify unusual actions that may indicate malicious intent. These tools can flag deviations from a user's typical behaviour and alert security teams to potential threats.

# 4. Response Protocols

### ▤ Incident Response Plan

Develop and regularly test incident response plans tailored for OT insider threats. These plans should include clear steps for detecting, containing, and recovering from insider threat incidents. Testing these plans ensures that your team is prepared in the event of an actual attack.

### ▤ Forensic Investigation

Conduct thorough forensic analysis to assess the extent of insider activities, identify vulnerabilities, and determine how the threat bypassed security. A detailed investigation will help uncover the methods used by insiders and allow your team to strengthen defenses moving forward.

### ▤ Legal and Compliance Procedures

Work with legal teams to ensure compliance with relevant regulations. Pursue legal action when necessary to hold malicious insiders accountable. It's important to understand regulatory obligations and data breach notification requirements to comply with industry laws.

# 5. Collaboration and Reporting

### ▤ Interdepartmental Collaboration

Foster close collaboration between IT, OT, security teams, and HR for effective threat management. Ensure all teams are aligned on detection and response strategies. Collaboration helps bridge the gap between operational technology and IT security, ensuring comprehensive threat mitigation.

### ▤ Timely Reporting

Implement clear and quick reporting mechanisms to escalate insider threats immediately. Ensure that all team members know the procedures for reporting and responding to potential insider threats, reducing the time it takes to act once a threat is detected.

# 6. Continuous Improvement

### ▤ Post-Incident Analysis

Conduct a thorough review after each incident to identify lessons learned, improve detection methods, and update response protocols accordingly.

These reviews should be used as a foundation for improving OT security processes.

■ **Regular Assessments**

Regularly assess and update security measures to adapt to evolving insider threat tactics and new vulnerabilities in OT environments. Security measures should evolve in tandem with the growing complexity of OT systems and the increasing sophistication of insider threats.

OT insider threats require a proactive, multi-layered approach to safeguard critical infrastructure. By implementing robust detection, mitigation, and response strategies, organisations can significantly mitigate the risks posed by insider threats in OT environments.

# POCKET GUIDE 4: CYBER CRISIS COMMUNICATION MAPS

Cyber Crisis Communication Maps are tailored tools designed for navigating and responding effectively to cyber security crises. These maps offer a structured framework to guide communication strategies during cyber incidents.

## 1. Purpose of Cyber Crisis Communication Maps

■ **Strategic Cyber Guidance**

Provide a visual guide for understanding the flow of communication during a cyber crisis. Clarify roles, responsibilities, and communication channels specific to cyber security challenges to ensure alignment across teams.

■ **Cyber Proactive Planning**

Facilitate proactive planning by identifying potential cyber crisis scenarios and communication strategies. This enables a swift, coordinated response to minimise reputational damage in the cyber security context and reduce confusion during an incident.

## 2. Key Components of Cyber Crisis Communication Maps

■ **Stakeholder Identification**

Identify and categorise key internal and external cyber security stakeholders. Tailor communication plans to address the unique needs and preferences of each stakeholder group to ensure the right information reaches the right people.

■ **Communication Channels**

Map out primary and secondary communication channels relevant to cyber incidents. Consider traditional media, cyber security-specific platforms, internal channels, and direct outreach. This ensures no communication gaps during a crisis.

■ **Decision-Making Protocols**

Define decision-making protocols specific to cyber crises to streamline communication approvals. Specify the cyber security incident chain of command and individuals responsible for key decisions to avoid delays and confusion.

## 3. Cyber Crisis Scenarios and Message Templates

■ **Scenario Mapping**

Anticipate various cyber crisis scenarios, such as data breaches, ransomware attacks, and system compromises. Tailor communication plans for each cyber scenario, accounting for unique challenges and audiences.

■ **Message Templates**

Develop pre-approved message templates for different cyber crisis scenarios. Ensure consistency in messaging while allowing for necessary adaptations based on the nature of the cyber security incident. Templates help maintain clarity and reduce reaction time.

## 4. Cyber Crisis Team Roles and Responsibilities

■ **Team Structure**

Define the structure of the cyber crisis communication team, including cyber security experts and communication specialists. Assign roles and responsibilities to team members based on their expertise in cyber security communication, ensuring efficient and effective handling of the crisis.

■ **Training and Drills**

Conduct regular training sessions and cyber crisis communication drills to familiarise the team with the communication map. Ensure team members are well prepared to execute their roles effectively in the dynamic cyber crisis environment, maintaining a cohesive approach.

# 5. External and Internal Cyber Communication Coordination

■ **Coordination Protocols**

Establish protocols for coordinating external and internal cyber communication efforts. Ensure alignment between messages directed at the public and those intended for internal cyber security stakeholders, minimising conflicting information.

■ **Real-Time Cyber Updates**

Implement procedures for real-time updates to keep all stakeholders, including cyber security experts, informed. Emphasise the importance of consistent and timely communication during cyber crises to prevent misunderstandings and misinformation.

# 6. Integration with Other Cyber Crisis Response Plans

■ **Collaboration with Other Cyber Teams**

Ensure alignment with other cyber crisis response plans, such as IT, legal, and incident response. Collaboration ensures all teams address the broader spectrum of cyber security challenges during a crisis, working toward unified crisis management.

■ **Continuous Improvement in Cyber Response**

Regularly review and update Cyber Crisis Communication Maps based on cyber incident feedback and lessons learned. Incorporate improvements to enhance responsiveness and adapt to evolving cyber threats. Continuous updates help ensure that your communication strategies remain relevant and effective.

# 7. Digital Communication Strategies in Cyber Crises

■ **Social Media Protocols**

Develop cyber-specific protocols for managing social media during a cyber crisis. Outline strategies for addressing misinformation and engaging with the online cyber security community to control the narrative and prevent panic.

■ **Digital Monitoring in Cyber Crises**

Integrate digital monitoring tools specifically tailored for cyber incidents. Monitor online sentiment and adjust cyber communication approaches accordingly to address cyber security concerns and mitigate damage to the organisation's reputation.

Cyber Crisis Communication Maps are indispensable tools for navigating the complexities of cyber security incidents. A well-developed map ensures a coordinated and strategic cyber response, fostering resilience and maintaining trust in the face of challenging cyber situations.

# POCKET GUIDE 5: MEDIA TRAINING FOR CYBER C-LEVELS

Media training equips Cyber C-Levels with essential communication skills for handling cyber security incidents effectively. This training enhances organisational resilience and reinforces public trust by ensuring clear and consistent messaging during crises.

## 1. Purpose of Media Training for Cyber C-Levels

■ **Rapid Cyber Response**: Enables Cyber C-Levels to respond promptly to cyber incidents, reducing reputational damage.

■ **Clear, Accurate Communication**: Ensures clarity and accuracy in communication with media, stakeholders, and the public.

■ **Trust & Transparency**: Builds trust through transparency and effective leadership during high-stakes cyber security situations.

## 2. Key Components of Media Training

**Message Development**

■ Create concise, consistent messages aligned with the organisation's cyber security stance.

■ Anticipate potential questions and prepare responses to maintain narrative control.

**Crisis Communication Techniques**

■ Train C-Levels in crisis-specific communication strategies to manage media interactions effectively.

■ Address the "no comment" challenge by transforming it into an opportunity for transparency.

**Media Interview Simulation**

■ Conduct simulated interviews to recreate high-pressure scenarios.

■ Provide constructive feedback to refine poise and message delivery.

# 3. Translating Jargon

### Simplify Technical Language

■ Train C-Levels to translate technical cyber security details into accessible language.

### Avoid Jargon Overload

■ Emphasise the need to avoid overly technical terms that may alienate or confuse the audience. Avoid cryptic phrases like "zero-day exploit" and instead describe it as "an unknown software flaw that hackers can exploit before it's fixed".

■ Use relatable analogies to clarify complex concepts for non-technical stakeholders. For example, explain a "firewall" as "a security checkpoint that inspects and controls data entering or leaving your network, like customs officers checking luggage at an airport".

# 4. Social Media Management for Cyber Incidents

### Effective Platform Usage

■ Educate executives on using social media for controlled, consistent messaging.

■ Ensure messaging consistency across social media channels to reinforce credibility.

### Response to Viral Incidents

■ Establish guidelines for addressing incidents that gain traction online.

■ Balance rapid responses with thoughtful messaging to maintain trust and credibility.

## 5. Building a Unified Communication Front

### Team Coordination

- Highlight the importance of aligned communication across the C-Suite team.
- Foster a unified front to project consistency and strength in messaging.

### Internal Communication Alignment

- Guide C-Levels in coordinating with internal teams to align internal and external messaging during crises.

## 6. Handling Sensitive Questions in Media Interactions

### Preparation for Difficult Queries

- Equip C-Levels with strategies for addressing challenging or sensitive questions calmly and confidently.

### Legal Constraints

- Provide guidance on navigating legal boundaries while remaining transparent.
- Collaborate with legal teams to ensure a harmonised approach to sensitive topics.

## 7. Continuous Improvement of Media Training

### Post-Incident Review

- Conduct reviews to evaluate communication effectiveness following a cyber incident.
- Identify areas for improvement and incorporate lessons learned into future training.

### Ongoing Education

- Encourage C-Levels to stay informed on emerging cyber security threats and evolving communication strategies.

  Media training for Cyber C-Levels is essential for fostering organisational resilience and ensuring clear, confident responses to cyber security incidents. Well-prepared executives strengthen public trust and reinforce the organisation's credibility during times of crisis.

# POCKET GUIDE 6: MANAGING DEEPFAKE RISKS

Deepfakes are AI-generated media that manipulate images, videos, or audio to fabricate events or portray individuals inaccurately. Understanding and managing deepfake risks is essential for protecting individuals and organisations from reputational and security threats.

## 1. Understanding Deepfakes

- **Definition**: AI-generated synthetic media that manipulates or fabricates content to create deceptive scenarios.
- **Technology Behind Deepfakes**: Utilises advanced machine learning, particularly Generative Adversarial Networks (GANs), to synthesise realistic yet fraudulent content.

## 2. Types of Deepfakes

- **Face Swapping**: Replaces faces in videos to create convincing yet false visuals.
- **Voice Cloning**: Mimics voices to produce authentic-sounding but altered speech.
- **Context Manipulation**: Alters context within media to misrepresent events and fabricate narratives.

## 3. Detecting Deepfakes

- **Digital Forensics Tools**: Use specialised software to identify anomalies or alterations in media content.
- **AI-Driven Solutions**: Implement algorithms to detect patterns unique to deepfake creation techniques.
- **Human Expertise**: Engage forensic analysts or content verification experts to identify subtle inconsistencies.

## 4. Mitigating Deepfake Risks

- **Public Awareness Campaigns**: Educate the public on deepfake risks and how to identify them.

- **Media Authentication Techniques**: Use watermarking, cryptographic signatures, and other methods to verify content authenticity.
- **Regulations and Policies**: Support and enforce policies that regulate the creation and distribution of deepfakes, especially in sensitive areas.

## 5. Real-Time Response Strategies

- **Content Removal and Flagging**: Rapidly remove and flag identified deepfakes on platforms to minimise their impact.
- **Disseminating Corrective Information**: Share verified information through trusted sources to counteract deepfake influence.
- **Platform and ISP Collaboration**: Work with online platforms and ISPs to ensure quick takedown of deepfake content.
- **Public Awareness Alerts**: Issue alerts to inform the public about deepfakes and their potential harm.
- **Legal Action and Reporting**: Report harmful deepfakes to authorities for potential legal action.
- **Media Engagement**: Collaborate with media outlets to prevent the spread of false information and encourage responsible reporting.
- **Continuous Monitoring**: Maintain monitoring systems to detect new deepfakes and enable swift responses.

## 6. Ethical Considerations

- **Consent and Privacy**: Emphasise respect for consent and privacy rights in the creation and sharing of media content.
- **Media Integrity**: Advocate for the ethical use of AI, discouraging the malicious use of deepfake technology.

## 7. Continuous Research and Development

- **Innovative Detection Solutions**: Support ongoing research to enhance detection technologies and counter evolving deepfake tactics.
- **Collaboration and Knowledge Sharing**: Foster partnerships among researchers, technology firms, and policymakers to address deepfake threats collectively.

A well-rounded approach involving technology, public awareness, regulation, and ethical responsibility is essential for managing the risks of deepfakes and maintaining trust in digital media.

# AFTERWORD

As we close this journey through *The Cyber Security Handbook*, it's worth reflecting on a crucial element that has underpinned much of what we've discussed: the role of crisis communications in the broader context of cyber security. If there's one takeaway from these pages, it's this: a successful technical response to a cyber incident is only half the battle. Even the most proficient investigation and remediation efforts can be undermined by an ineffective communications strategy.

Cyber security is no longer a purely technical domain; it's a business issue. The narratives that arise from cyber incidents – whether through media, social platforms, or the threat actors' own dark web leak sites – can shape the perceptions of customers, shareholders, employees, regulators, and the public at large. Every communication during a cyber crisis is an opportunity to build trust, demonstrate leadership, and mitigate long-term reputational harm. Conversely, missteps in communication can exacerbate the damage, erode confidence, and undermine an organisation's technical response to a breach.

One of the most important lessons from this book is the need for a communication response tailored to the unique challenges of each incident. The dynamic nature of modern cyber attacks – threat actors releasing data on the dark web, extorting employees directly, or engaging other direct and indirect extortion tactics – demands that organisations remain nimble, proactive, and deliberate in their messaging. Silence is not a shield; it is an invitation for others to control the narrative, as this book describes so aptly in Chapter 6.

As highlighted throughout this handbook, effective crisis communications must be deeply integrated into an organisation's incident response framework. This means aligning messaging with the technical investigation, ensuring that what is shared externally and internally is consistent, clear, and reflective of the known facts. It means simplifying technical complexities into accessible language that empowers stakeholders to act decisively and responsibly. Above all, it means responding with transparency and integrity to build and sustain trust.

The ecosystem of stakeholders in a cyber crisis is vast and varied. From employees and customers to regulators, shareholders, and law enforcement, each group has its own priorities and concerns. Successful crisis communications require a nuanced understanding of these audiences, delivering tailored messages that address their specific needs while maintaining consistency across the board. This book has provided valuable strategies for navigating these complexities, ensuring that no stakeholder is left uncertain or uninformed.

As adversaries continue to evolve their tactics, public-facing extortion campaigns have become an increasingly common weapon in their arsenal. These tactics seek to exploit fear, confusion, and mistrust among key stakeholder groups. Organisations must respond to these public threats with equal measure, providing timely and factual updates to counter misinformation and reassure those affected. In this context, the importance of preparing for public-domain responses cannot be overstated.

Crisis communications is not merely a defensive tool – it's a critical part of your response playbook; an opportunity to demonstrate resilience and leadership. The principles and practices outlined in this handbook emphasise that organisations can emerge stronger from a cyber incident, provided they approach their communications with care, foresight, and purpose. By communicating effectively, you're not only addressing the immediate crisis but also setting the stage for long-term recovery and renewed stakeholder confidence.

As you move forward, let this book serve as both a guide and a reminder: in the realm of cyber security, every word, every message, and every action matters. By embracing the role of communications as a cornerstone of your cyber strategy, you equip your organisation to face whatever challenges may come – not with fear, but with clarity, confidence, and resolve.

Thank you for investing your time in this critical topic. The insights and strategies you've gained here are not just tools for navigating crises; they are a call to action. In the ever-evolving world of cyber threats, preparation and communication are the keys to resilience and success.

BEN HOCKMAN
Cyber Crisis Response & Communications Expert

# BIBLIOGRAPHY/WEBOGRAPHY

## Articles

Arghire, I., "Iran-Linked hackers use 'Mia Ash' honey trap to compromise targets", www.securityweek.com, 1 August 2017.

Holtom, B. H., Edmondson, A. C., and Niu, D., "5 Tips for communicating with employees during a crisis", *Harvard Business Review*, 9 July 2020.

Bruce, I., Tran, K., Shey, H., and Burn, J., "Don't wait for a crisis to act", Forrester Blog, https://www.forrester.com, 23 July 2024.

Budel, S., "The precision medicine revolution will be driven by diagnostic technologies", *Forbes*, 28 April 2020.

Deloitte, "Silence the noise", *Deloitte Review*, Issue 24, January 2019.

Meyer, M. L., Masten, C. L., Ma, Y., Wang, C., Shi, Z., Eisenberger, N. I., and Han, S., "Empathy for the social suffering of friends and strangers recruits distinct patterns of brain activation", *Social Cognitive and Affective Neuroscience*, vol. 8, no 4, 2013, p. 446–454.

Gill, M., "In times of crisis, empathy fuels loyalty", Forrester Blog, https://www.forrester.com, 18 December 2020.

Hatch J., "Defending the digital world: The growing cyber threats", BAE Systems, https://www.baesystems.com/en/digital/blog/defending-the-digital-world, 26 June 2017.

Henderson, C., "Bringing threat intelligence and adversary insights to the forefront: X-Force Research Hub", IBM, https://securityintelligence.com/x-force, 3 August 2023.

Sheridan, K., "85% of data breaches involve human interaction: Verizon DBIR", www.darkreading.com, 13 May 2021.

UK government, "UK and Singapore sign new innovative digital trade deal", press release, https://www.gov.uk, 25 February 2022.

## Books

Billois, G. (with Nicolas Cougot), *Cyberattaques. Les dessous d'une menace mondiale*, Hachette, 2022 (only available in French for now).

Coombs, W. T., and al., "Situational crisis communication theory (SCCT) and application in dealing with complex, challenging, and recurring crises", in Jin,

Y., and al., *Advancing Crisis Communication Effectiveness: Integrating Public Relations Scholarship with Practice*, Routledge, November 2020.

Fung, A., and al., *Full Disclosure: The Perils and Promise of Transparency*, Cambridge University Press, 2007.

# Surveys, reports and studies

Allianz, "Cyber perils outrank Covid-19 and broken supply chains as top global business risk", Allianz Risk Barometer, www.allianz.com, 18 January 2022.

Bank of England, "CBEST thematic", 2023.

CrowdStrike, "Global Threat Report", 2021.

Deloitte, "Cyber crisis management: readiness, response, and recovery", https://www2.deloitte.com/content/dam/Deloitte/ch/Documents/audit/ch-en-cyber-crisis-management.pdf, 21 April 2016.

Deloitte, "Detect & Respond: Staying ahead of cyber threats", www.deloitte.com/cyber, 2021.

ENISA, "Cybersecurity Threats Fast-Forward 2030: Fasten your security-belt before the ride!", 11 November 2022, and "Identifying security threats and challenges for 2030", www.enisa.europa.eu, March 2023.

Gartner, "The state of digital transformation for financial services business-line leaders", www.gartner.com, 2019.

GlobalData, "Nearly half of global consumers are influenced by changes in society when purchasing products, says GlobalData", www.globaldata.com, 24 March 2022.

IBM, "Majority of global C-suite executives are rapidly accelerating digital transformation due to COVID-19 pandemic, but people and talent are key to future progress", IBM Global C-Suite Study, IBM Newsroom, 30 September 2020.

IBM, "Threat intelligence index", 2022.

IDC, "Future enterprise resiliency and spending survey", www.idc.com, March 2022.

Immersive Labs, "Immersive Labs Global Study finds improved response time to threats, yet resilience efforts still fall short", https://www.immersivelabs.com, 2 August 2023.

KnowBe4, "2021 Cybersecurity awareness training benchmarks".

Lenaerts-Bergmans, B., "Dark web monitoring", www.crowdstrike.com, 27 April 2023.

Levy I., "Cybersecurity and behaviour: Beyond technical defences", NCSC.

Marsh, "Top risks for UK businesses revealed", https://www.marsh.com, 1 November 2023.

Mordor Intelligence, "Content Delivery Network Market Size & Share Analysis –
Growth Trends & Forecasts Analysis (2024–2029)", https://www.
mordorintelligence.com, 2023.

OECD, "Of bytes and trade: Quantifying the impact of digitalisation on trade",
OECD Trade Policy Paper, www.oecd.org, no 273, May 2023.

Ponemon Institute, "State of AI in cybersecurity", 2024 Report.

Ponemon Institute, "The 2017 Cost of Data Breach Study: Your response plan is key"
2017.

PwC, "Crisis communication", PwC's Global Centre for Crisis and Resilience,
www.pwc.com.

PwC, "Four fault lines show a fracturing among global consumers", Report: PwC's
Global Consumer Insights Pulse March 2021, www.pwc.com.au, June 2021.

Tech Nation, "Lifting the lid on how UK tech boomed in 2020", 2021 Report,
https://growth.technation.io, 23 November 2021.

UK government, "UK Cybercrime Cost Report", https://www.gov.uk, 2011.

Wavestone, "How to create Crisis Comms that will take your crisis response to
the next level", https://wwa.wavestone.com, 2024.

World Economic Forum (WEF), "Global Risk Report", https://www.weforum.org,
January 2023.

# Index

**P** Pearson

# JOIN THE PEARSON BUSINESS BOOK CLUB

> FREE Monthly Webinars with expert authors to help boost your personal and professional development

**Discover Now** ↘